Praise for Anthony C. Caputo's *Digital Video Surveillance and Security*

"A must read for any security professional contemplating based security system from an author who was instrun America's largest city's homeland security systems."

—Jeffrey Menken, Video and Wireless Network IT Architect, IBM

"The subject matter of this book will appeal greatly to security practitioners and small business or home owners looking for guidance to install IP-based surveillance cameras and integrated systems."

—Brian Baker, CPP, Brian D. Baker Security Group, State College, PA

"This book covers quite a bit of information in great detail and is a great 'how-to' reference. The illustrations are good and as we all know, a picture tells a thousand words, so that is a good aid.... A fine manual and description of Digital Video Surveillance."

—Roger J. Martinez, President and CEO, Quantum Crossings, LLC, Technology and Telecommunications

"A valuable resource of real-world, in-depth knowledge on networked video surveillance systems."

—Bo Larsson, CEO, Firetide, Inc.

Digital Video Surveillance and Security

Digital Video Surveillance and Security

Anthony C. Caputo

ELSEVIER

AMSTERDAM • BOSTON • HEIDELBERG • LONDON
NEW YORK • OXFORD • PARIS • SAN DIEGO
SAN FRANCISCO • SINGAPORE • SYDNEY • TOKYO

Butterworth-Heinemann is an imprint of Elsevier

Butterworth-Heinemann is an imprint of Elsevier
30 Corporate Drive, Suite 400, Burlington, MA 01803, USA
The Boulevard, Langford Lane, Kidlington, Oxford, OX5 1GB, UK

Notices
Knowledge and best practice in this field are constantly changing. As new research and experience broaden our understanding, changes in research methods, professional practices, or medical treatment may become necessary.

Practitioners and researchers must always rely on their own experience and knowledge in evaluating and using any information, methods, compounds, or experiments described herein. In using such information or methods they should be mindful of their own safety and the safety of others, including parties for whom they have a professional responsibility.

To the fullest extent of the law, neither the Publisher nor the authors, contributors, or editors, assume any liability for any injury and/or damage to persons or property as a matter of product liability, negligence or otherwise, or from any use or operation of any methods, products, instructions, or ideas contained in the material herein.

Library of Congress Cataloging-in-Publication Data
Application submitted

British Library Cataloguing-in-Publication Data
A catalogue record for this book is available from the British Library.

ISBN: 978-1-85617-747-4

For information on all Butterworth–Heinemann publications
visit our Web site at www.elsevierdirect.com

Printed in the United States of America

10 11 12 13 10 9 8 7 6 5 4 3 2 1

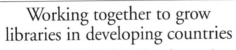

This book is dedicated to my family and they all know who they are.

Contents

Preface   xiii

Acknowledgments   xv

About the Author   xvii

1. Introduction to Digital Video Security   1
   Introduction   1
   Closed Circuit Television   3
   Big Brother Is in the Restroom   6
   Digital Video Security   9
   General Security   11
   Case Studies   13
   Chapter Lessons   16

PART I   CHOOSING THE RIGHT EQUIPMENT

2. Digital Video Overview   19
   Introduction   19
   Analog to Digital   20
   Analog Versus Digital   21
   Worldwide Video Standards   22
   Interlaced Lines   23
   Progressive Scanning   23
   Resolution   25
   Digital Color Depth   25
   The Wonderful World of Pixels   26
   Digital Video Surveillance Resolutions   31

Digital Video Formats   31

MPEG   33

Analog Camera and Digital Video Encoder Versus the IP Camera   36

Chapter Lessons   38

3. Digital Video Hardware   39

The Evolution of Video Surveillance Hardware   39

How Cameras Work   39

Choosing the Right Cameras for the Right Job   45

Configuring Digital Video Encoders and IP Cameras   63

Digital Video Cables and Connectors   71

DVS Troubleshooting   77

Chapter Lessons   88

4. Understanding Networks and Networked Video   89

Introduction   89

The Power of the Network   89

Getting Wired   91

Why Ethernet   93

Setting up a Star Network   103

Bandwidth   107

VLAN   107

Video Networking   108

Networked Video Delivery Methods   109

Understanding Broadcast and Multicast Packets   113

Remote Access – Your Home away from Home   115

Lessons Learned   115

Chapter Lessons   122

5. Wireless Networked Video   **123**

    Introduction   123

    Introduction to RF   123

    Without Wires?   123

    Radio Frequency   124

    Access Point   126

    Antennas   131

    WLAN Standards   138

    Wireless Security Options and Considerations   148

    Channel Planning   150

    Configuring Access Point Radios   150

    Configuring a Mesh Radio   151

    Wireless Antenna Coaxial Connectors   155

    Antenna Coaxial Cables   155

    Wireless Troubleshooting   160

    Chapter Lessons   167

PART 2   APPROACHING THE PROJECT

6. Site Survey   **171**

    Introduction   171

    License Plate Recognition   173

    Human Recognition   173

    Power = Camera, No Power = No Camera   177

    Camera/Video Site Survey   180

    Network Infrastructure Site Survey   183

    Wireless Site Survey   187

    Chapter Lessons   194

7. Choosing the Right Software   195
    Video Management System Software   195
    Chapter Lessons   218

8. DVS Archiving and Storage   219
    Introduction   219
    DVS VMS Requirements   219
    The Anatomy of a Computer   222
    Client/Server Architecture   223
    Upgrading Hardware for DVS   224
    The Network Operating System   237
    IP Cameras   239
    Network Accessibility   239
    Troubleshooting   242
    Chapter Lessons   249

9. Project Implementation   251
    Introduction   251
    Project Management Institute and the Real World   251
    Chapter Lessons   280

10. Security Integration and Access Management   281
    Security Integration   281
    Electronic Access Control and Management   294
    Troubleshooting   302
    Chapter Lessons   305

Appendix   307
Index   315

Preface

One could say that my career in technology has followed the convergence of our digital world. I've been working with digital video, with multimedia, since the mid-1990s. My first experience with networking communications was an experiment in 1989 with a pre-press company who wished to test a new method of delivering digital desktop publishing pages over a 2400-baud modem.

My extensive experience in digital video resolution and compression (codecs), bandwidth, and streaming within a video editing environment and my years in a multitude of networking technologies including writing the *Build Your Own Server* book and working in streaming media for Warner Bros., BMG Music, and commercial training applications made it easy to understand digital video surveillance. The security aspect was a natural transition from my days as a system analyst, when everything was about business process improvement.

In early 2006, I joined the IBM Project Management Office for the City of Chicago's Operation Virtual Shield (OVS), a city-wide homeland security initiative that linked a multitude of cameras through the downtown Chicago area. My history in digital video, and networking technologies and as an author and writer, coupled with my experience as a certified Project Management Institute (PMI) Project Management Professional (PMP) and instructor, made me a valuable addition to the team. My initial responsibility was to write the procedure manuals for the system. As many technical writers experience when working on projects with the complexity of something such as OVS, there are very few individuals who know everything about all aspects of the system and those people are usually too busy to sit down on a regular basis to assist in the documentation of how the system works. Bits and pieces can be accumulated from various project managers, architects, designers, implementers, contractors, etc., so in the end I went out into the field to observe and record.

Sometimes figuring out how something is supposed to work and then fixing it is the only way to find out how it *really* works, so you can accurately document it. I stuck my nose and knowledge in too many places and wasn't allowed to leave the field again.

This book is the accumulation of years of hands-on experience in the field, and I believe it severs the ties that video surveillance has had for decades with closed-circuit television (CCTV). The world of analog television has also jumped into the digital age, thanks to the convergence of digital entertainment and communications, and video surveillance has jumped along with it.

Digital Video Surveillance and Security explains the concepts that are becoming the new standards in video security, both theoretically and from the field where the only days your hands aren't dirty from handling the real tools and equipment are the days when you're wearing gloves to protect your hands from the subzero temperatures.

Anthony C. Caputo
2009

Acknowledgments

Special thanks to everyone at IBM who gave me the opportunity to thrive: Roger Rehayem, James Sara, Timothy Herlihy, Jeffrey Menken, Dave Bisset, Ted Gary, Jim Lautenbach, Jodi Samsa, and Ladislao Delgado. Grand thanks and appreciation for suggestions, contributions, and comments: Dave Bisset, George Shapkarov, Jimmy Jimenez, Jeffrey Menken, Roger Rehayem, Terry Hennessy, Tim Roudebush, Mike Intag, Murali Repakula, Sivakumar Kailias, Pramod Akkarachittor, Joey Gerodias, and Michael Lane. Special thanks to my friends at MPEA for their help and support: Vince Gavin, Ellen Barry, Rich Piotrowski, and Tony Clark. Special thanks to my friends at the City of Chicago's 911 Center, Office of Emergency Management and Communications (OEMC) for their help and support: Aric Roush, and Joe Zito. And also thanks to: Gina Fritts, Brandon Ballschmeide, Donny Rutkowski, Rob Fleenor, Roger Martinez, Neil Salkind, Ed Guy, Pam Chester, Nancy Hoffmann, and Megan Berry.

About the Author

Anthony C. Caputo is currently working for IBM Global Technology Services on the Digital Video Surveillance Homeland Security Projects for the City of Chicago, Chicago Housing Authority, and the Metropolitan Pier and Expo Authority (MPEA). He has written articles, business plans, and books about the business benefits of technology for 22 years, including McGraw-Hill's *Build Your Own Server*, and has presented at conferences on the importance of a network security plan when introducing the Internet into any organization. He has 10 years of experience as a successful entrepreneur in both the entertainment and technology arena and has helped build five companies (three in technology) within the past 15 years. He's a certified and subject matter expert in a number of technology disciplines including project  management with PMI (PMP), networking technologies and architecture, Genetec Omnicast Digital Video Surveillance software, Firetide Mesh Network Engineer, object-oriented analysis and design for business process improvement, and a Microsoft Certified Professional. He holds a Certification as an IBM e-business Solution Advisor, helping IBM write the exam for e-business advisor certification.

1
Introduction to Digital Video Security

Introduction

Visual surveillance began in the late nineteenth century to assist prison officials in the discovery of escape methods. It wasn't until the mid-twentieth century that surveillance expanded to include the security of property and people. The astronomical cost of these first security camera systems, based on traditional silver-based photographic cameras and film, limited their use to government buildings, banks, and casinos. If questionable activity was discovered, the monitoring security firm would develop the films in a secure, private darkroom laboratory to analyze at a later date. Live television was occasionally used during special events to monitor a crowd, but law enforcement was usually limited to the television studio to view the multiple cameras.

The theory behind visual surveillance was founded on the same four key factors that are still prevalent today. These are

1. Deterrence
2. Efficiency
3. Capable guardian
4. Detection

Deterrence

If potential criminals are aware of the possibility of being watched and recorded, they may determine that the risk of detection far outweighs the benefits. Visual surveillance as a deterrent is used from casinos to retail settings to public transportation. Countries all over the world use video surveillance, focusing its use mostly on public transportation (planes, trains, and autos) and select public areas. Based on an Urban Eye study (www.urbaneye.net), 86% of these international installations are for the prevention and detection of theft, and 39% also serve as a deterrent of violent crime. The amount of crime prevented by using video surveillance is based on the environment and whether the system is solely passive, active, or both. A passive system uses video recordings after an incident to help solve a crime. An active system is monitored by security personnel who are dispatched at a moment's notice. Historically, the most effective crime prevention video surveillance systems do more than record crime in the background. One dramatic example is Chicago's Farragut High School, a public school notorious for its major acts of violence, locker thefts, and vandalism, all of which nearly disappeared within a year after the installation of a closed circuit television (CCTV) surveillance system. Many American cities have likewise seen a reduction in crime due to the addition of a video surveillance implementation and strategy.

In a recent UK Home Office Research Study on the effectiveness of video surveillance as a crime deterrent, 46 surveys were done within public areas and public housing in the United States and the United Kingdom. Of the 46 studies, only 22 had enough valid data to be deemed acceptable for publication. All 22 published surveys showed significant reduction (as much as 50%) in burglaries, vehicle theft, and violent crimes (see detailed report at www.homeoffice.gov.uk). However, it's rather difficult to analyze data on the effectiveness of video surveillance systems due to the many variables in the complexity of the areas of coverage and general displacement. For example, the decrease of crime within an area monitored by video surveillance cameras may have forced criminals to move to a different location, thus displacing the violent crimes. Enclosed areas of coverage – such as parking garages and lots, buildings, and campuses – have better success with video surveillance than large outdoor areas as long as there's a clear presence of a "capable guardian," which can be increased police or security guards or the electronic eyes of security cameras.

Efficiency

Reviewing video surveillance footage at the same time as watching live surveillance provides additional information about a situation, allowing users to make better decisions about deploying the right kinds and numbers of resources. Depending on the number of security cameras and their location, this simultaneous viewing of live and archived video can confirm a sleight of hand or any illegal activity before a patron, customer, or suspect is approached by a security force. In 2007, the Dallas, Texas, Police Department used video footage from 559 incidents to assist in 159 arrests. Their experience indicated that a single police officer monitoring live and archived video can cover far more area than a field officer, including usable image captures of license plates from 300 yards away.

Capable Guardian

In the article "Social Change and Crime Rate Trends: A Routine Activity Approach" by Lawrence Cohen and Marcus Felson (*American Sociological Review*), the authors suggest that crime prevention includes the presence of a "capable" supervising guardian. That guardian doesn't have to be present, just watching. Today, the guardian doesn't even have to be watching, just archiving using smarter technology. Current video surveillance includes sophisticated video analytics software with the capability of monitoring areas for programmable situations (e.g., bookmark all red automobiles) such as abandoned cars or backpacks, circling vehicles, or even specific license plates. Video analytics can upgrade an originally passive security system into an active one. This introduces the capable guardian by giving the passive surveillance system a "brain" and allowing it to be more responsive to potential criminal activity.

Detection

Detection is the higher profile success factor, providing tangible evidence that video surveillance works. Britain is well known for its video surveillance system, providing law enforcement with the ability to follow anyone throughout the city of London through the use of over 200,000 cameras (with over 4 million cameras throughout the country). This system helped locate four London-born terrorists including the well-publicized CCTV images of suicide bomber Hasib Hussain. Likewise, the arrests of Jon Venables and Robert Thompson in the high-profile British murder case of James Bulger were directly linked to images reviewed on the surveillance system. Furthermore, Scotland Yard convicted 500 criminals using their CCTV database that included 3 years of data on 7000 offenders.

Closed Circuit Television

CCTV, which uses traditional radio frequency (RF) technology, rather than photographic technology, was introduced in the 1980s and provided a more cost-effective and real-time method of video surveillance.

Fake Cameras and the "False Sense of Security" Liability

There are many options in video surveillance, all of which feed the desire to take advantage of the deterrent factor. There are a number of companies marketing fake video surveillance cameras, which in the short run may initially help deter criminal activity, but even a fake camera, although a deterrent to criminal activity, implies "security." People walking through the loading docks of a store may believe they're safe because they see cameras and assume a security force is watching. If a criminal incident happens and the cameras are fake, that false sense of security may provide the basis for a winning lawsuit in today's courts. Even though the criminal broke the law and/or trespassed on private property, a court may fault a company for installing fake security cameras. This could be considered breach of contract, for knowingly stating that there was security when there wasn't; negligence, for lulling the employee with a false sense of security; or failure to heed police recommendation if an incident happened in the same area in the past. Ultimately, fake cameras could cost more in legal fees and settlements than installation of true video surveillance. Based on the Video Surveillance Guide Web site (www.video-surveillance-guide.com), once cameras are identified as fake they have been known to increase criminal activity, sometimes with devastating consequences.

Today's concept of video surveillance has its roots in the analog world of television. The framework of CCTV is a simple one, using the same analog signal you'd receive from your old pre-digital television. A single camera monitors one place and sends it to a CRT television monitor at another place using a coaxial cable. Usually the system has a single command center where security personnel watch black-and-white and/or color monitors of various cameras. Multiplexing technology provided the ability to watch more than one camera on a single monitor, or automate a cycling of various camera feeds on a single monitor to expand the area of coverage. While it's true that many security professionals and companies still use CCTV and the concept of a centralized "command center," not everyone has the space, money, and/or resources for such a system. The idea of wiring a house, office, building, or campus with coax cables from every camera to a control unit and then to each CRT monitor is costly, time-consuming, and thanks to internetworking technologies, unnecessary.

Figure 1-1 depicts an example of a CCTV installation that monitors select areas of coverage. The first installation was designed and developed for the separate parking facility. This implementation included several fixed position cameras on each floor of the parking garage, stairwells, and exits, all connected directly to a primary control unit (PCU) for management of each video stream. The PCU is a simple device for managing the input and output of video feeds through the coax cables. Ancillary utilities and devices can provide simple integration of some alarms, but this technology has limited capabilities and a complex integration process.

A single monitor in the parking garage management office was connected to the output for monitoring cameras. The PCU offered shuffling of each camera feed and select intervals and a keyboard to input the call number for each camera, or the ability to scroll through each camera one by one.

Several years later the campus was expanded; unfortunately the previous CCTV installation wasn't designed to extend the system into other buildings. An underground site survey uncovered various fiber, Ethernet, and power connectivity, but the conduit was either full or damaged over time. Feeding new runs of coax required trenching and/or boring to replace poor conduit runs between said buildings, thus the plan to run coax cables (for video) between the parking facility and the main building was abandoned due to cost. Another isolated CCTV system was designed and developed within the main building. These cameras were installed inside loading docks, exits and entrances, main entrances, and service corridors. A new model camera was introduced into this system that required more connectivity than coax cables for video. Many pan-tilt-zoom (PTZ) cameras were installed, requiring separate wiring interconnectivity with the new PCU for camera controls using a proprietary protocol.

In addition to this phase of the expansion, a primary command center was built to house a new security office with a CCTV control console for several CRT monitors to view the cameras and a new model keyboard with a built-in joystick to access and control the new PTZ cameras.

Closed Circuit Television (CCTV)

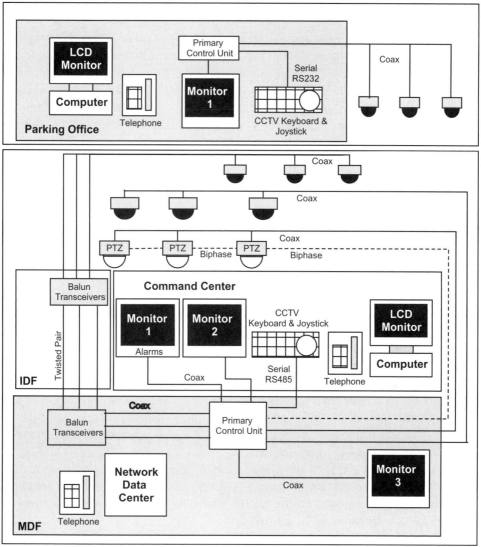

FIGURE 1-1 A typical CCTV topology.

The main building didn't have any existing conduit or spare conduit pathways to run coax throughout it and into the new buildings on campus. Video Balun transceivers were used to transfer the coax video signal to existing telephone twisted-pair wires between locations, as each building was interconnected to each telephone interim distribution facility (IDF), or a secured closet with twisted-pair terminals for the telephones and a network switch for the computers. The plain old telephone service (POTS) lines were linked into each IDF, the main distribution facility (MDF), the command center, and any room with a telephone. Although the use of Balun can affect the video quality, it made possible the installation of dozens of cameras that were originally deemed too costly.

Multiple monitors were installed within the new command center with a single monitor assigned to alarm displays. Once any alarm system integrated into the PCU was activated, the nearest camera to that location would be displayed on that particular monitor. The chance that the fixed camera would be pointed in the direction of the incident depends on the initial requirements for area of coverage. If there are three select emergency exits, panic alarms, or door sensors, a single fixed camera can only watch one. This creates a one in three chance of catching a specific incident on video (either live or recorded). The command center also included a computer designated for filing incident reports online. These can be accessed by management personnel at a later date from a database via the computer network.

Big Brother Is in the Restroom

Twenty-five years ago general and business communications were primarily synchronous. To accomplish almost anything there needed to be someone on the other side of the table or telephone line, especially when dealing with national and international business. The same holds true for security and video surveillance. Typically, CCTV is most effective as an active system with a security guard monitoring the corridors with someone else, somewhere, watching video surveillance monitors for support. Everything is synchronized and everything happens in real time.

Today everything moves quickly and technology has added to life's complexity with the magic of fax machines, computers, PDAs, and mobile phones – all capable of delivering multimedia through multiple channels to a mass of recipients, without having to synchronize with anyone. The message is received and the response happens when it happens. The world has changed dramatically since the invention of the analog television over a century ago, yet it's only been recently that television has caught up with our faster mail, faster computers, faster networks, and faster foods. This information overload has forced an asynchronous world where there's too much information and not enough time and resources to absorb it.

When I speak about my work with IBM and the City of Chicago homeland security, I typically get the "Big Brother is watching" comment, referring to George Orwell's omnipresent socialistic watcher called Big Brother in his book *1984*. The "Orwellian" fear is that someone, somewhere may be watching you through a video surveillance camera this very minute, and that someone, somewhere is abusing the system to violate your freedom and privacy. Fortunately, you're not doing anything illegal – just reading this book, and hopefully not behind the wheel of a getaway car, idling in front of a neighborhood bank or terrorist target, or speeding through a red light at a high-risk intersection. Yet, if you're not doing anything wrong, you shouldn't worry and there's actually more of a threat from spyware or a Trojan virus on your home or business computer, watching your every move on your desktop, reading your files, and looking through digital photos and movies of your family, than the chance of a video surveillance camera watching your every physical move. While video surveillance monitors the linear world in real time, spyware is software that lives in virtual nanoseconds with computer farms multi-tasking trillions of computations against millions of unsuspecting data targets in the blink of an eye, making digital data and transactions far more vulnerable and a truly unnerving threat. (Tip: Turn off the computer when you're not using it.)

The security systems I've designed and implemented are strictly for monitoring activity of a potential target, whether from a terrorist threat or neighborhood thieves or vandals. The systems are deployed for the protection of lives and property, not to watch individuals as they jaywalk across the street. Albert Alschuler, a law professor at Northwestern University, makes the point that public camera systems don't violate any privacy concerns because they're installed in public locations. They have been implemented for public safety and crime prevention and aren't for the exploitation of individuals. An individual with a camcorder or the paparazzi may have a more self-satisfying agenda, and thanks to the Internet can violate privacy laws with far more devastating consequences than video surveillance cameras, which are kept secured and in strict confidence.

Digital video security (DVS) works because humans can't be everywhere, all the time. In today's world very few businesses can afford to have human resources sitting behind a desk 24/7, watching television monitors. Such an active approach would also be very limited when dealing with a large crowd. There are too many cameras and too few eyes watching them, but once you plug your video and security systems into the ever-growing telecommunication infrastructure, DVS becomes a very cost-effective and simple replacement for CCTV. It's this complete convergence between entertainment and telecommunications that makes DVS possible. It provides a meaningful solution to video surveillance because you don't have to be actively monitoring an area of coverage to achieve your goals. DVS archives footage using automated software and when a programmed event occurs (e.g., motion, door sensor, vandalism), the DVS system sends an alarm using the asynchronous ways that are already part of today's business infrastructure.

Convergence

DVS is made possible by the proliferation of standardized broadband telecommunication technologies and video compression formats. Once the data pipe reached a level to actively handle heavier loads of data, coupled with the introduction of better video compression quality, algorithms, and processing power, it was possible to deliver hefty video streams from place to place without standardized television signals, DVD players, or game consoles. DVS works much the same way as YouTube, making it possible for anyone, anywhere to receive and view a single stream of video without any compatibility concerns. Only DVS isn't about entertaining or training, but the protection of life and property.

Universal access is made possible by the ubiquitous nature of digital data tele-communications, which can provide cable television, broadband Internet connection, and telephone on a single pipe, while a decade ago it was only a pipe dream (no pun intended). This technological interconnectivity is brought to you by "convergence." Once elusive and now a reality, convergence is the integration of the separate technologies that run our daily lives. This new universal delivery method makes it possible to watch movies or listen to music on the television, DVD player, game console, computer, and mobile phone (Figure 1-2).

As the world turns digital – from analog television to HDTV, telephones to voice over IP (VoIP) and cell phones, analog magnetic security access cards to smart cards, and VCRs to DVRs or PVRs – it's no longer a struggle to integrate these separate systems. Once they're digital, there's no need to have a single access point for live video or recorded surveillance footage because DVS shares pertinent information as well as sends immediate alerts to management, security, and/or the local police department using the already existing information technology infrastructure in place from security access management to cable modems to cell phones.

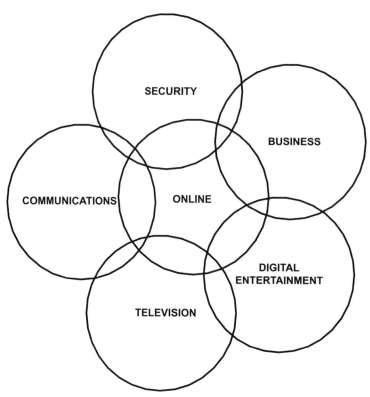

FIGURE 1-2 Convergence of our digital world.

Digital Video Security

DVS is intelligent networked digital video surveillance: the integration of all new and existing security assets including cameras, alarms, networks, archiving, and analytics, with managed accessibility to monitor those assets from anywhere, at any time. It builds upon and/or extends current surveillance, security, and IT infrastructure and investments.

DVS is made possible by the interconnection and digitization of the world. The digital video formats and the delivery method ignore the traditional analog video methodologies. Digital video isn't RF waveforms (as with traditional analog video either through the air or coax cable), but a structured universal data file of zeroes and ones. There's no National Television System(s) Committee (NTSC) or phase alternating line (PAL) in the delivery mechanism (see Chapter 2), only in its creation and only if it's required in the delivery method (NTSC or PAL), such as for a DVD player (but even many of the newer DVD players no longer need the differentiation). However, unlike analog video or television which has been around for over a century, DVS needs serious computing power and the better the imagery, the higher the power requirements (see Chapter 2). The microprocessor, or central processing unit (CPU), has reached a level of speed and power that can generate high-quality live and recorded video using the same sensors inside today's more sophisticated digital camcorders (see Chapter 3), which a decade ago cost thousands of dollars and are now available in high definition for a few hundred dollars.

Figure 1-3 depicts a CCTV example converted into a complete DVS solution. Our previous site surveys uncovered one simple fact that opened the door to a number of possibilities, all of which will encapsulate the entire system under one video management system (VMS). The difference between the PCU, from the CCTV example, and the VMS is integration and interface; more software and less hardware. While your PCU outputs video to traditional CRT analog monitors, the VMS accomplishes this on a computer desktop, laptop, or server because it's already set up for digital feed. That includes any traditional CRT PC monitor or a widescreen HDTV LCD wall-mounted monitor. VMS speaks the language of digital, making integration with other digital systems more accessible.

Key to this convergence is the existing Ethernet network that interlocks all the buildings on campus using highly efficient fiber optic cable and/or copper wire (Ethernet). The existing analog cameras, both fixed and PTZ, are plugged into a video encoder (using the existing coax for video and twisted pair for PTZ) converting the analog signal into digital video capable of transmitting through the existing computer network. New network switches are introduced to isolate the DVS system from Internet and intranet traffic, and these switches ride the fiber and/or copper between facilities through a virtual local area network (VLAN). VMS software is installed onto a new, dedicated VMS server, and the client software is installed on the select desktop computers allowing secured access into this new DVS network.

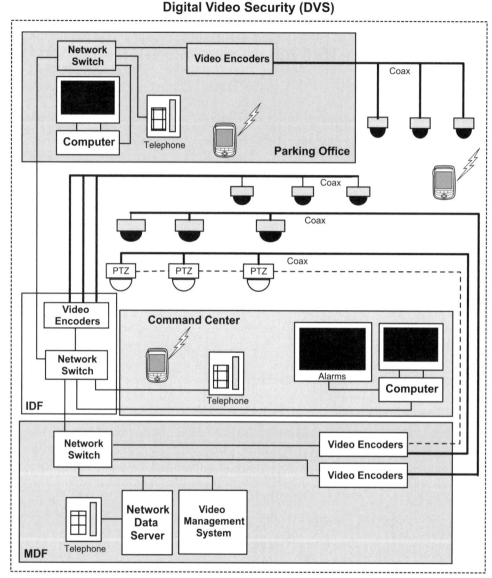

FIGURE 1-3 The typical CCTV topology, converted to DVS.

The VMS software also offers event messaging and alarms. The POTS telephones and twisted-pair lines are replaced by a new VoIP teleconferencing system which, being on the same Ethernet network, can now deliver messaging to the telephone LCD screens – both land lines and mobile phones. The video surveillance system is now smarter. *This* is digital video security.

General Security

An important point to remember is that DVS doesn't replace your alarm system, keys, deadbolts, encryption, or unique login authentication. It enhances it by allowing you to physically see what's happening from anywhere online. Recording 30 days of video of your property is pointless if someone can break in and steal or damage your DVR or reach up and cut the video feed (Figure 1-4).

General security rules still apply. Never underestimate the need to protect your new DVS investment with alarm systems, environmental benefits, and even conduits to protect cables from exposure to the elements and/or wire cutters. Fiber optics are the most vulnerable, which, unlike copper wire, can actually break like glass.

Anyone who has designed and implemented security cameras and/or systems will tell you there's no such thing as 100% security or a vandal-proof security camera and system. There may be "vandal-resistant" camera models, which include high-impact enclosures with security screws (screws with a unique head requiring a special tool) and even bulletproof enclosures, but nothing says "vandal-resistant" like inaccessibility. Inaccessible doesn't mean installing cameras and other security devices (e.g., sensors, audio alarms, and wireless communication devices) at the corner of the roof on top of a building with roof access. Disabling any security camera is simple. All you need is a can of spray paint, or for those who are really out of reach (on top of poles that are 20-foot plus), your average paintball gun. The key to vandal resistance is making it as difficult as possible to deactivate the security system, including the cameras, and implementing enough cameras programmed to react to these situations and work together to find the culprits (more on this in Chapter 10). (See Figure 1-5.)

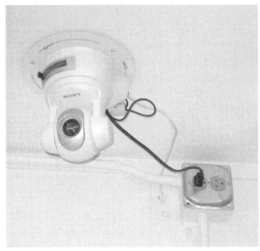

FIGURE 1-5 Not only was this camera accessible and easily destroyed, but the IP address of the network was written inside the NEMA box for all to see (not shown here).

FIGURE 1-4 Avoid exposing power and data cables, providing an easy method of disabling the security camera.

This book includes details about designing and implementing a DVS system, but making sure the system stays intact is key. The core foundation for vandal resistance is the inaccessibility of not only the cameras, but their data feeds and power supply as well. What's the use of installing a 35-foot monolithic pole when a simple crescent wrench can open the hand-hole at the base of the pole and cut the power and data feeds? There are poles designed for high-profile crime areas with hand-holes 15–20 feet above ground, hopefully making them inaccessible to criminals (see Figure 1-6).

When using poles, hand-holes are a necessity, not only for installation, but also for maintenance, and the location of those hand-holes should be based on the environment (see Figure 1-6). High-profile, secure locations may use embedded ground hand-holes. Typically, power is fed from somewhere other than where the data is received and transmitted, depending on distance. If power is accessible nearby, but the environment isn't as secure, then go the distance and opt for security and avoid power daisy-chains whenever possible. Most outdoor lighting poles are daisy-chained for power between poles, because they require synchronization to an electric timer or photocell. This works well for lights, but security systems require "uninterrupted power" (more on this later). New power lines can be fed, using the existing conduit (if not damaged or crushed) or new conduit (as long as it's not exposed and within reach as depicted in Figure 1-7). But the daisy-chain creates many more points of failure because the power is terminated at the hand-hole of each pole, making each security camera on the chain vulnerable (see Chapter 6 for more information).

FIGURE 1-6 Keep power and data access out of reach.

FIGURE 1-7 Exposed conduit is an easy target.

Case Studies

Navy Pier, Chicago (Figure 1-8)

Historic Navy Pier is located on Chicago's lakefront. It's the Midwest's top tourist and leisure destination, attracting more than 8.6 million visitors a year. Navy Pier reaches outward into Lake Michigan for three-quarters of a mile and is completely surrounded by water on three sides. IBM's network video monitoring system implementations for Navy Pier included 128 new high-resolution Axis IP cameras that were installed in and around Navy Pier. In addition, IBM integrated approximately 100 existing analog cameras, including Pelco, Sony, and Panasonic on four separate undocumented monitoring systems and two separate alarm systems with over 300 alarm functions. The installation and configuration of video analytics provided Navy Pier, the Office of Emergency Management and Communications (OEMC), and the Chicago Police Department with a smart camera programmed to monitor and control boat traffic within the north inlet. IBM also created map interfaces (see Figure 1-9) of all Navy Pier levels, giving even newly hired security personnel intuitive and immediate recognition of each camera and alarm location.

There were two old but functional alarm systems integrated. The panic alarms, located in the garage facilities, provide patrons with an emergency call button and siren. Upon activation, the closest PTZ cameras (up to three) pan and zoom to the target location and populate special armed tiles within the software. The video can be a live feed and/or a pre-recorded playback to review the location seconds before the specific alarm activation, and the Command Center is instantly alerted with audio, color-coded lighting, e-mail, and messaging signals. The Smith Museum of Stained Glass Windows includes silent parameter and pressure-sensitive alarms for the protection of the museum exhibits

FIGURE 1-8 Chicago's Navy Pier along the lakefront *(photo courtesy Navy Pier and the MPEA).*

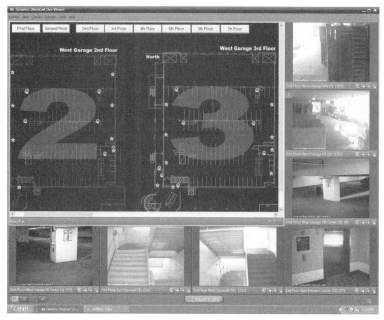

FIGURE 1-9 Maps of each floor were added to help locate the nearest camera.

and informs the personnel in the Command Center in the same manner as the panic alarms.

Navy Pier's new Security Command Center provides multiple officers with their own workstations with large screen displays each featuring review and control of up to 32 cameras simultaneously (Figure 1-10).

FIGURE 1-10 New Command Center *(illustration courtesy Jim Jimenez, IBM).*

All new displays, multiple workstations, lighting, climate control, and the unique audio and color-coded alarms for panic alarms, the Museum of Stained Glass Windows, and the existing fire alarm systems were integrated within the limited space available to create a state-of-the-art facility. Official training classes, complete as-built documentation, and mentoring were part of the skills transfer process, but what really made the transition successful was the intuitive interface (more on software in Chapter 7). The system is designed to help the security personnel monitor the huge facility and provide access to 30 days of archived video footage. Even as the system was implemented, the cameras sighted a thief running off with the money in a vendor's cash drawer.

Marquette Photo Supply, Chicago

Marquette Photo Supply is a privately owned camera store located in Chicago since 1947. It's located in a neighborhood that has seen many changes over the years (Figure 1-11).

Unfortunately, the neighborhood has been inundated with vandalism and robberies. Many businesses in the area have come and gone with the changes, but those that have endured thwart crime by using window bars, silent alarms, and high illumination for clear visibility of the entry and rear service entrances and exits. After a brief site survey to uncover the areas of coverage and power, only four cameras were required for optimum effectiveness. One was installed in the rear, covering the rear entrance. The remaining three were up front within the store showroom, monitoring the side exit, front entrance, and cash register. These cameras needed to be wired and have infrared capabilities to monitor activity 24/7, especially during off hours. Assessing if the desktop computer in the back office had the adequate requirements to be converted into a digital video recorder (see Chapter 8) was also part of the site survey. The computer was unable to be converted so it was upgraded with more processing power, additional memory, and a second hard drive to store the video archives for the required two weeks.

FIGURE 1-11 Marquette Photo Supply store *(used with permission by Joseph Herbert, owner).*

Not only do the cameras deter crime, but they create a visual record of every transaction for a two-week period, with those transactions now easily accessible through the digital archives and software. This endeavor costs less than $1000, considerably less than the Navy Pier project. Today, thanks to DVS, there's a video surveillance solution for everyone.

Chapter Lessons

At the end of most chapters in this book will be a troubleshooting section specifically focused on the chapter's contents and some real-world lessons learned. The term "best practices" is too broad a concept as each project is different and the only best practice is gathering a clear list of specific requirements: how and when the system will be used, by whom, and the area of coverage. Each project is dramatically different to the point that creating templates can be difficult.

Choosing the Right Equipment

2

Digital Video Overview

Introduction

The difference between digital video and standard analog video is the method of delivery. Digital video follows the laws of all digital technology; breaking down video imagery into a binary data stream composed of ones and zeros delivered as low-voltage electronic pulses over copper wire or light pulses over fiber optics. Anything that requires digital churning of bits (*bi*nary dig*its*) uses a microprocessor and stores information in binary "on" and "off" (ones and zeroes).

Digital video isn't necessarily better than the traditional analog signal, but it's far more consistent in its delivery and transmission and takes far less bandwidth with the right digital compression. Analog video is a waveform, which inherently can be corrupted and/or changed by environmental, atmospheric, or technological interference. In fact, because any waveform is allowable and inconsistent, there can be difficulty differentiating noise and distortion from the core analog signal. People who still use the old "rabbit-ears" antennas, in the days of analog television, often experience this interference. Digital video, however, is a mathematical absolute, using more effective compression techniques to deliver more data, including parameters for redundancy, retries, and quality. Digital video is also somewhat universal, thanks to the proliferation of the Internet. Unlike analog video, which has several standards, with two predominant international standards (National Television Systems Committee; NTSC and phase alternating line; PAL) and limited radio frequency (RF) spectrum for delivery, digital video is only limited by bandwidth.

In Japan, 20 MHz of RF bandwidth is used for analog television. This is over three times that of the United States, and that bigger RF pipe provided Japan with the means to deliver high-definition television (HDTV) over their traditional analog channels of video communication. The United States is limited to only 6 MHz in an RF spectrum crowded by a multitude of applications. The United States needed to turn to digital technology to fit more data within that same bandwidth (ten digital channels fit into the same bandwidth as a single analog channel), hence the need to switch to digital technology for HDTV television video transmissions (see Figure 2-1). However, digital transmissions aren't limited exclusively to video or even a single stream of video. As mentioned in Chapter 1, the convergence of all things digital can create a multimedia interface similar to the Internet, but now delivered through television. Based on a study by the Television Bureau of Advertising (www.tvb.org), there are over 115 million

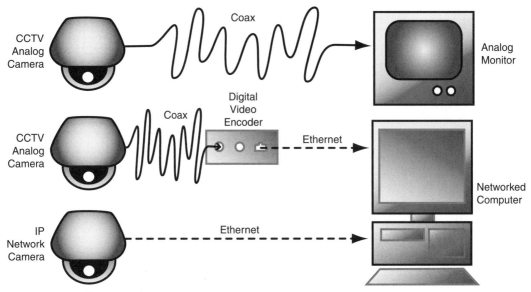

FIGURE 2-1 Digital versus analog.

U.S. households with a television (or 98% of the U.S. population), while only 64% of U.S. households own a computer and use the Internet, and within that group only 42% use broadband (the rest are still on 56 K narrowband dial-up). Digital television can provide the advantages of the power of the network to the 34% without the means to access online information, and broadband connectivity to the 58% still struggling with a POTS dial-up connection. To paraphrase John Barrett, Director of Research, Park Associates (a research firm for digital living), in an *Information Week* interview, "without access to the Internet, you are economically disadvantaged."

Analog to Digital

It's easy to understand the profound change in the way we communicate when we observe how video transmissions have changed within the past decade. Prior to the digital network (and Internet), video delivery and communications were limited to the same standards developed almost a century ago. These standards used the Federal Communications Commission (FCC) regulated RF analog signal to secure and protect transmissions from being hijacked or corrupted before making it to your television. This same signal was encapsulated using magnetic tape for videocassette recorders and later digitized for DVD. However, video wasn't the first analog media that made the leap into a digital delivery media. Music was originally released as "record albums" – the same method of delivery invented by Thomas Edison in 1877, and in 1911 the original recorded cylinders were replaced with recorded discs (thanks to Columbia Record's victrolas). It wasn't until much later in the twentieth century that those vinyl record discs had any competition – first with 8-track tapes (1970s) and then later cassette tapes (1980s). Although the compact disc (CD) player was introduced in 1982, the theory behind it can be traced back to British engineer Alec Reeves who patented "pulse code wave modulation" (PCM) in 1943. PCM is the *digital representation of an analog signal* and is used today when converting any waveform into mathematical data, including video.

The quality difference between a cassette tape and CD is the same as a videocassette tape and a DVD. Simply put, when digitizing the analog signal the process also includes filters and quantification to produce the best possible delivery. Although the music industry first introduced the digital alternative to a long antiquated analog system, the benefits included far more than performance quality. Disc players – whether they're CD players or DVD players – have fewer moving parts as well as digital printed integrated circuit boards, which have a longer life span than analog circuits and components.

Analog Versus Digital

The image on any monitor or screen, whether the traditional analog cathode-ray tubes (CRT) television or computer monitor, is the accumulation of rectangular dots called pixels (an abbreviated version of picture element). All images, video or static, are measured in pixels. In the analog world, each of these pixels is composed of three color dots of red, green, and blue (RGB). The naked eye blends these three color dots into a single color on the phosphor CRT screen. The phosphor emits light in direct proportion to the intensity of the electron beam hitting the screen. Analog television screens have a color depth of about 256 levels for each of the three colored layers, so each analog pixel has a spectral range of about 16.8 million colors (16,777,216). The same holds true for LCD monitors, only the result is achieved differently. The LCD monitor is a thin, flat display of liquid crystals sandwiched between two pieces of polarized glass called the "substrate." A small fluorescent bulb (the backlight) illuminates the first substrate, and when the electric current from a thin-film transistor activates the crystals to align at various levels of light and color, it passes through the second substrate, creating what you see on the screen.

While the CRT monitor projects the RGB dots using phosphor, the LCD monitor still needs those same three degrees of colored dots to create its own pixel. Ideally, these three dots or layers would be in exactly the same spot, but they're close enough and overlaid at varying percentages to fool the naked eye into seeing 16.8 million different colors. The printed color page, under magnification, also shows a similar methodology using the four-color process of CMYK (cyan, magenta, yellow, and black). These four simple colors, used in varying percentages, can closely simulate (enough for the naked human eye) the same 16.8 million different colors.

Worldwide Video Standards

There are three major worldwide analog television standards: the American NTSC color television system, the European PAL, and the French-Former Soviet Union Sequential Couleur avec Memoire (SECAM).

■ ■ ■ ━━

Why 29.97 fps?

Believe it or not, television began in black and white with 525 lines scanned in 1/30 of a second. This produced a line scan rate of 15,750 Hz (525 × 30), but with the invention of color television room was needed for the new RGB signals and so the 15,750 Hz became 15,734 Hz, changing the 30 frames per second (fps) to 29.97 fps (15,734/525).

━━ ■ ■ ■

The primary difference between the three systems is the vertical lines and frame rate (see Table 2-1). While NTSC uses 525 lines, both PAL and SECAM use 625 lines. NTSC frame rates are slightly less than half of the 60-Hz power line frequency, while PAL and SECAM frame rates are exactly half of the 50-Hz power line frequency.

The standard NTSC analog signal is sent via select frequencies between 30 Hz and 6 MHz, originally (and still today) through the airwaves to rabbit-ears antennas or over coax cable. NTSC is a standard 525 lines of 720 pixels for a total of 378,000 pixels per frame and an analog signal that delivers a complete image 29.97 fps. PAL delivers 625 lines of 720 pixels at 25 fps or 450,000 pixels per frame. This is a mature technology and although it suffers from the lack of quality control inherently built within digital technology, analog television transmissions are limited (in the United States) to 6 MHz. Depending on the age of the television set and/or monitor and the method of transmission, analog television transmissions end up displaying about 330 lines (WebTV® systems use a 544 × 372 pixel screen size for a standard analog television screen). However, when that analog signal is captured for digital transmission, the digital encoder can "see" all 11,302,200 pixels per second to encode because of the shorter distance between the analog camera and the digital encoder. Once digitized, the analog signal improves by taking the allowable waveform signal and converting it to a digital absolute, filtering out noise and focusing on the core signal. The only limit is the resolution, and unlike a CRT television monitor that's fixed at a single resolution, the typical computer display (CRT of LCD) provides many resolution options that depend on how the display is used.

Table 2-1 Worldwide Television Standards

	Lines	Active Lines	Vertical Resolution	Horizontal Resolution	Frame Rate
NTSC	525	484	242	427	29.97
PAL	625	575	290	425	25
SECAM	625	575	290	465	25

Interlaced Lines

Analog video, including television, VHS, and analog camcorders, uses an "interlaced" method of scanning (see Figure 2-2). Interlacing works on lower resolution analog televisions and monitors because it lacks sufficient detail to become visible. The interlaced scan-based images were developed for CRT television monitor displays, splitting the number of visible horizontal lines across a standard TV screen into two separate odd and even fields that alternately refresh 29.97 fps. Any slight delay in refresh rates between the two groups creates a jagged or crosshatch distortion (see Figure 2-3). This is because one line group refreshes at a slower rate than the other.

Progressive Scanning

Digital encoders provide a deinterlacing filter, or progressive scanning option, to avoid the jaggedness from interlaced video (which is a throwback to the original analog video). Deinterlacing or progressive scanning (see Figure 2-4) requires more processing power and is especially important at the highest resolutions because the distortion can be more easily recognized. All cameras with slower shutter speeds can create blurred images of motion. This is emphasized even more with interlaced distortion because the camera isn't fast enough to capture the motion.

A progressive scan, as opposed to an interlaced one, scans the entire picture line by line in sequential order every 1/16 of a second without segregating the scene into two separate fields. Digital technology doesn't need interlace scanning for video presentation. Digital video eliminates the "flickering" effect as long as the display monitor is also of suitable quality, and more important, the digital devices provide the ability for deinterlacing and/or progressive scanning.

Most DVD players now employ progressive scan at 480p, as the majority of homes have a standard television set limited to the 480i resolution. If a DVD movie is displayed

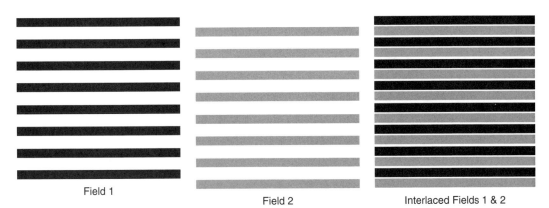

Field 1   Field 2   Interlaced Fields 1 & 2

FIGURE 2-2 Interlaced diagram.

FIGURE 2-3 Interlacing – notice that only the moving objects are affected.

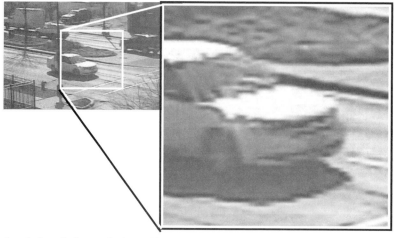

FIGURE 2-4 Deinterlacing eliminates the jagged lines.

on an HDTV, the DVD player can "upconvert" or "upsample" the 480p to the maximum resolution on the HDTV (e.g., 1080i). Depending on the DVD player and HDTV, this can be an automated process or something that needs to be manually accomplished.

Therein lies the difference between viewing HDTV on a 1080i (interlaced scan) display versus a 1080p (progressive scan) display. Traditional analog television has a maximum of 480 scan lines, which is the maximum output for analog video. Digital cable or satellite television offers high-definition 1080 progressive scan resolution, but only with an HDTV. If the HDTV, computer monitor (CRT or LCD), or projector is limited in color depth, resolution (less than 1080), and/or progressive scanning capabilities, it will convert the 1080p into 1080 interlaced scan lines.

Resolution

While the standard analog television format uses vertical rectangular pixels, the new HDTV format is composed of smaller square pixels (similar to computer monitors). There are four and a half times more digital pixels in the same space used for an analog NTSC pixel (8-bit vs. 32-bit), so it holds four and a half times more detail for a sharper image. All digital resolution is measured in these pixels and all digital imagery, including printed material, is pixel based. Although printing requires 300 dots per inch (dpi) for true color and quality, most digital monitors have either 72 or 96 dpi capabilities (CRT or LCD), so any onscreen digital display doesn't require the same resolution as the printed page to achieve the same visual quality. On an analog CRT television screen the resolution is fixed, so the dpi will decrease based on the size of the television screen. The pixel width of the television screen divided by the size of the screen determines its dpi depth. For example, an 8-inch-wide television would have a screen resolution of 560/8 or 70 dpi. Most televisions are 27 inches diagonally; thus the dpi resolution would only be about 26 dpi. You may have an excellent video signal, but the display may not be able to achieve that level of quality.

Digital Color Depth

In the digital world of absolute mathematical equations, pixels aren't measured in dots that vary in color by light-emitting phosphor, but in bits and bytes. The same 256 levels of color seen on any traditional color television include 16 or 24 bits per pixel in the digital world. Color resolution, also referred to as color depth, is the number of bits of data used to store information about the color of each pixel. A higher color resolution increases the range of colors and detail in the image (and increases the file size). Typical color resolutions are 8, 16, 24 (true color), or 32 bit.

Color depth can also determine the quality of the visuals presented onscreen. Color depth is merely the number of actual bits that make up a pixel. The original 8-bit color was only capable of 256 colors, but once that doubled to 16 bits onscreen color

Table 2-2 Digital Color Depth Is the Number of Bits per Pixel

Number of Bits	Number of Colors	Formula
8 bits	256 colors	2^8 $(2 \times 2 \times 2 \times 2 \times 2 \times 2 \times 2 \times 2)$
16 bits	65,536 colors	2^{16}
24 bits	16,777,215 colors	2^{24}

exponentially reached 65,536 colors (see Table 2-2). Eventually monitors displayed 24-bit color depth per pixel reaching the 16.8 million colors of color television, but with far more detail in dots per inch and four times more color pixels.

The majority of today's monitors allow for 16- and 24-bit color. The "32-bit color" is typically only 24-bit color with an additional 8 bits of non-color metadata, giving the operating system and high-end graphics card additional functionality to change resolutions or use dual monitor displays.

■ ■ ■ ▬▬

Tech Tip

Different software video players and hardware monitors may not have the built-in functionality to adapt to select color depths, instead showing as solid black or receiving an error message regarding codec incompatibility. It's good practice to change the color depth to 16, 24, or 32 bit to see if the video image comes up on screen, or to try another monitor before discarding it as corrupted.

▬▬ ■ ■ ■

The Wonderful World of Pixels

Table 2-3 shows many of the available digital video resolutions (there are select monitors that can display higher resolutions than HDTV) including the higher megapixel cameras. Keep in mind that as everything digital is measured in pixels, it's entwined between the resolution displayed and the display resolution. For example, as Figure 2-5 depicts, a 50-inch LCD HDTV monitor is capable of 1920 × 1080p. That's 1920 × 1080 pixels in progressive scan. A typical low-resolution security camera generates a 352 × 240 pixel (or CIF) image, a high-resolution security camera is 704 × 480 (or 4CIF), and DVD movies and D1 resolution are 720 × 480. The video surveillance software controls how the security camera video is displayed on an HDTV monitor, whereas DVD players move automatically to full screen. However, because the screen resolution is higher than the DVD resolution, it "upconverts" the 480p to fill the screen, but all that does is increase the size of the pixels. To simulate HDTV (using 1080i), upsampling is done. Upsampling is interpolating between the existing pixels to obtain an estimate of their values at the new pixel location. In other words, a DVD player, HDTV, projector, or

Table 2-3 The Wonderful World of Pixels

Display Standard	Pixel Settings	Aspect Ratio	Total Pixels per Frame	
SQCIF	128 × 96	4:3	12,228	
QCIF	176 × 144	4:3	25,344	
VHS (NTSC)	320 × 480	4:3	115,200	
CIF	352 × 240	4:3	76,800	
1/2 D1*	352 × 480	4:3	168.960	
2/3 D1*	464 × 480	4:3	222,720	
DCIF	528 × 320	16:9	168.960	
S-VHS (NTSC)	530 × 480	4:3	192,000	
VGA	640 × 480	4:3	307,200	
Standard NTSC TV 480i	640 × 480	4:3	153,600 × 2 (interlaced)	
2CIF*	704 × 240	4:3	168.960	
4CIF	704 × 480	4:3	337,920	
Standard NTSC DVD 480p	720 × 480	16:9	345,600	
D1	720 × 480	16:9	345,600	
SVGA	800 × 600	4:3	480,000	
XVGA	1024 × 768	4:3	786,432	
XVGA+	1152 × 864	4:3	995,328	
	1280 × 1024	4:3	1,310,720	1.3 megapixels
SVGA+	1400 × 1050	4:3	1,470,000	1.5 megapixels
	1600 × 1200	4:3	1.920,000	1.9 megapixels
WSVGA	1680 × 1050	16:10	1,764,000	1.8 megapixels
WUXGA	1920 × 1200	16:10	2,304,000	2.3 megapixels
QXGA	2048 × 1536	4:3	3,145,728	3 megapixels
HDTV 1080p	1920 × 1080	16:9	2,074,000	2 megapixels
HDTV 1080i	1920 × 1080	16:9	1,037,000	
HDTV 720p	1280 × 720	16:9	922,000	
HDTV 720i	1280 × 720	16:9	461,000 × 2 (interlaced)	

*These resolutions trick the naked eye by cutting the amount of horizontal lines in half.

whatever digital display device is used may have upsampling capabilities that do more than just stretch the 720 × 480 DVD movie to full screen. Upsampling actually calculates the pixel valleys and fills in the blanks to create a 1080i resolution.

There's a correlation between the pixels displayed, pixels that are generated to be displayed, and pixels printed for hard copy. Typically, in the movies, law enforcement captures an image of the bad guy on a CCTV camera and prints it out as an 8″ × 10″ glossy photograph for everyone to see (including the viewers). What they show is a crisp photograph assuming high-definition resolution. This isn't the magic of modern CCTV, but the magic of Hollywood. To print a high-quality image of a screen capture at a size you can see on screen, you'll need more than three times the number of pixels. In other words, a 32-bit, 2550 × 3300-pixel image is equal to an 8.5″ × 11″ photograph (at 300 dpi). Any CCTV camera converted to digital via an encoder still has a maximum output

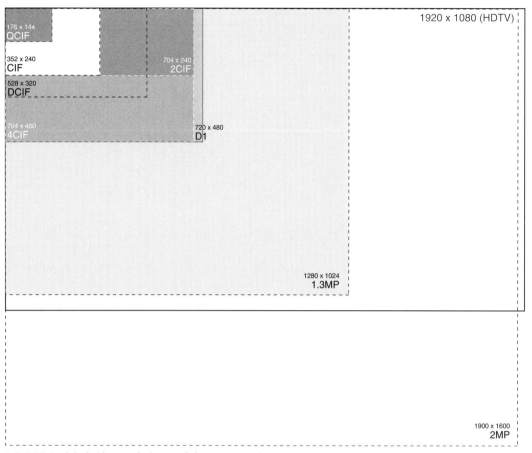

FIGURE 2-5 Digital video resolutions and the HDTV monitor.

of 32 bits, 704 × 480 pixels at only 96 dpi (max screen resolution). For that image to appear as a high-definition photograph, you would need to convert the 96 dpi to 300 dpi, but because there are only 337,920 pixels captured in the original frame, you're forced to change the proportions for printing resolution. At 300 dpi, the 704 × 480, or 7.3″ × 5″ onscreen, becomes a 2.4″ × 1.6″ photograph. This is the reason a print an image from the Web is either very blurry and pixilated or tiny. Figure 2-6 presents a unique visual example of how the traditional NTSC 4CIF resolution image compares to a megapixel image, while focusing on the same area of coverage. Most of the time manufacturers and solution providers present a wondrously scenic area of coverage because the megapixel cameras have more pixels, which means a larger digital image. What if the primary target is still within that 4CIF resolution? A combination of camera, telephoto lenses, and mounting can provide the same area of coverage, but with the superior resolution of the megapixel camera.

FIGURE 2-6 (Left) Printed screen shot of a 4CIF high-resolution analog video security camera connected to a digital video encoder (704 × 480). (Right) Printed screen shot of the same area of coverage, using a megapixel camera.

Upsampling Tech Tip

I've been working with digital video since 1994 and have noticed that 72 dpi onscreen versus 300 dpi on paper can be detrimental when trying to present printed imagery. Over the years I've worked with many image software applications and the finest by far is Adobe Photoshop. It's not cheap, but it has a multitude of functions and benefits that other applications lack. One of those features is the ability to convert, upsample, and manipulate digital imagery for print. Here is a great trick I've learned over the years: always work with images in RGB mode (not CMYK or indexed color) and you'll get the best results.

Step 1: Maximize Display Resolution

In Microsoft Windows, right-click on the computer desktop and choose PROPERTIES > SETTINGS > SCREEN RESOLUTION and make sure it's set to the maximum resolution. This is based on the equipped video graphics card and type of monitor. Some video graphics cards offer a larger resolution than the monitor's "native" resolution, but offers the ability to scroll around, off screen, and in the inset display. Choose the maximum.

Step 2: Full Screen Display

Take the paused digital video clip where you'd like to print out a still and, using the Video Management System (VMS) software (which may also have upsampling capabilities), maximize the image to full screen. It may appear blurry, but this will ultimately be corrected.

Step 3: Screen Capture

Unless the VMS application provides a means of capturing the entire video image display (not streaming), you'll have to find another application such as Snag-it, to do a full screen capture. You can also use the PRINT SCREEN feature built into Microsoft Windows. If you press and hold the left CRTL key, and then press PRTSC (Print Screen key), Windows captures the image onto the Clipboard (memory).

Continued

Step 4: Save Image

Launch Adobe Photoshop and choose FILE > NEW to open up a new art board image. Press OK as Photoshop knows what size the image in the Clipboard is and automatically creates the new image at the right size. Once the new (blank) image is open, press the left CRTL key and the "V" key to paste the full screen capture into your new art board. Choose LAYER > FLAT-TEN IMAGE to reduce the size and then choose FILE > SAVE AS. Type in the name of the file and the location where you'd like to save it and make sure to choose TIFF (*.TIF, *.TIFF) as the format. TIFF is a lossless compression to maximize the quality and is typically used in print.

Step 5: Convert for Print

Although you've captured the image, it's still in 72 dpi (or 96 dpi, depending on your monitor). To change it to print quality, choose IMAGE > IMAGE SIZE and take note of the file size and PIXEL DIMENSIONS. Once you change the RESOLUTION under DOCUMENT SIZE from 72 to 300 dpi, the file size will jump to over 20 times its original size. This is where the process becomes tricky, as typically, 20 times the file size is too much to upsample, so try choosing two to ten times the size and see how Photoshop handles that specific image for you (always keeping the resolution at 300 dpi).

What this accomplishes is simple. Let's say your digital video security (DVS) video is set to CIF, which is only 352 × 240 pixels in size. If your monitor is 1280 × 1024 pixels or your HDTV is 1920 × 1080, you've increased the size of the image to the full resolution of the monitor. However, that doesn't mean all video graphics cards, monitors, or VMS applications will upsample the image (add additional pixels to fill in the gaps). If any one or all of those accomplish this, you're one step ahead. Once you add the image into Photoshop, it takes care of it from there (see Figure 2-7).

Adobe Photoshop also has some helpful tools to improve the quality of the upsampled image. Choose FILTER on the top menu, and then choose SHARPEN. This will help minimize the pixilation and blurriness of the image.

FIGURE 2-7 The picture at the upper left is using the existing CIF bits (352 × 240) and the image on the lower left is using existing 4CIF bits (704 × 480) converted to 300 dpi (shrinks the image to about one-quarter the size on screen). The picture on the right shows the same frame capture, following the previous tips for upsampling using Adobe Photoshop CS3, an Nvidia 9500GT, and a 1-GB video graphics card on a 19-inch LCD monitor at 1280 × 1024 resolution.

Digital Video Surveillance Resolutions

Much like commuting in Chicago is measured in time rather than distance, streaming video is measured in bandwidth that determines the size of the video. The more pixels, the more required bandwidth, and there's no difference for digital video surveillance. The following labels are derived from video-conferencing technologies and multimedia applications, which were the premiere applications for streaming digital video. They follow the H.263 standards by the International Telecommunications Union (ITU).

Quarter Common Intermediate Format (QCIF; pronounced "KU-SIFF") is a standardized 176 × 144-pixel image format. Its smaller resolution makes it possible to stream through low-bandwidth environments, but doesn't provide sufficient detail for video surveillance beyond recognizing that there are specific objects present (e.g., auto, person).

Common Intermediate Format (CIF) is the most standardized viewing and recording size for digital video surveillance originating from an analog signal. At 352 × 288 pixels, it matches the quality of television and home video systems (VHS tapes). A CIF stream can be encoded and converted to a non-interlaced scan (either deinterlaced or progressive scan) and can use the maximum NTSC frame rate of 30 fps. CIF only uses half of the horizontal resolution of the PAL system at 288 lines. Many digital encoders also offer a 2CIF option, which eliminates one of the two interlaced fields and turns 480 horizontal lines into 240 lines.

Four times Common Intermediate Format (4CIF), with a maximum resolution of 704 × 480 pixels, is typically the highest resolution video originating from an analog camera (remember the NTSC stream from analog video camera is 640 × 480 pixels). Table 2-4 shows the different formats related to size on an LCD monitor.

Sixteen times Common Intermediate Format (16CIF), with a 1408 × 1152-pixel resolution, is rarely used for anything but wall-mounted video presentations.

Digital Video Formats

Digitized video began with the proliferation of the PC and the CD-ROM drive or what was called the "multimedia computer," which became mainstream in the mid-1990s when the number of CD-ROM-equipped computers reached 50 million consumers.

Table 2-4   Traditional Video Surveillance Formats

Picture Format	Luminance Pixels	Luminance Lines	Uncompressed 30 fps (Mbps)	
			Gray	Color
SQCIF	128	96	3.0	4.4
QCIF	176	144	6.1	9.1
CIF	352	288	24.3	36.5
4CIF	704	576	97.3	146.0
16CIF	1408	1152	389.3	583.9

Although digital video began on the Apple computer with QuickTime.MOV video, Microsoft quickly entered the market with their proprietary digital video solution. The audio video interleave (AVI), still in use today, is also accompanied by the more streaming friendly Windows Media Audio (.wma), Windows Media Video (.wmv or .wm), and the Advanced Systems Format (.asf). It's important to understand these file formats and extensions because any of these specific formats will play on most all Windows operating systems machines using the built-in Windows Media Player. Video compression technologies are known as "codecs," a term derived from the *co*mpression and *de*compression methodology. There are a few pertinent reasons why digital video and compression/decompression are typically synonymous. The primary reason is size: both bandwidth and storage space. Digital video files, depending on the quality, can be huge. The original CCIR 601, using interlaced NTSC video at 720 × 576, with 50 fps generated a 165-Mbps data rate. That's just the data rate per second, not the file size. Since the introduction of the CCIR 601, technology has developed various methods of compressing, decompressing, and playing digital video.

Digital imagery is made possible by compression technologies. These unique algorithms create a sequence of steps for converting any image into digital data and at the same time compressing it to its smallest file size, then decompressing it for presentation with minimal loss of quality. All active digital image formats (static or motion) are based on international standards set by the International Organization for Standardization (ISO) and International Electro-technical Commission (IEC). They're also globally accepted by the ITU as the de facto standard for digital still picture and video coding.

In the mid-1980s, a group called the Joint Photographic Experts Group (aka JPEG, pronounced "Jay-PEG") was formed to develop a foundation for digital image compression, and the group's first public release (in 1991) was the JPEG standard. The majority of images used online and in multimedia use JPEG (or the revised JPEG 2000) for compression and decompression.

This compression scheme is "lossy" (loses data), as it dumps actual repetitive bits of information. Obviously, you can't just discard information, so the JPEG algorithm divides the image into squares that you can actually see in poorly compressed JPEG files. This orderliness is accomplished (once again) by way of discrete cosine transformation to turn the squares of data into a set of curves in various sizes, which are then connected to create an image. At this point the formula is programmed to discard fragments of the file. The reduction requirements are taken into consideration when the algorithm deletes the less significant aspects of the image data (the smaller curves), those which contribute less to the overall "shape" of the image. The JPEG compression can generate unwanted side effects, such as false color, blurs, and blockiness if taken to an extreme.

The real issue with using the JPEG algorithm isn't its lossy codec, but how it's used as motion JPEG (MJPEG). MJPEG is the digital video version of the JPEG compression family and follows the compression algorithm used in online images. MJPEG is a higher quality video codec, because it forces the use of all required frames per second. MJPEG video compression is lossy, dropping details from the image that may be hidden in solid black or white areas, but recording at 30 fps isn't enough to shrink the video stream to a more manageable size.

Unlike MPEG-4, which includes a specific formula of I- (key frame), P-, and B-frames to creation motion, MJPEG uses every frame as if it were an animation. Every frame is a key frame, so the file sizes for MJPEG are much larger and the required bandwidth is bigger than a comparable MPEG-4 file.

MPEG

The Moving Picture Experts Group (MPEG) was formed to create a standard in digital video after developing a number of versions of the MPEG video codec. Although the method of compression is very similar to JPEG, the method of delivery is quite different, shrinking the bandwidth and storage space down to one-quarter that of motion JPEG:

- MPEG-1 (ISO/IEC 11172) was first released in 1993 and is still used in many multimedia applications today.
- MPEG-2 (ISO/IEC 13818) is the standard used for DVD movies, delivering data at a whopping 6 MB per second, making it too cumbersome for video surveillance.
- MPEG-3, or the more common MP3, is used as an audio and music codec.
- MPEG-4 (ISO/IEC 14496) has become the standard for streaming digital video (recorded or live), because it provides flexibility and considers the bandwidth and storage requirements. MPEG-4 can deliver 6 MB per second like MPEG-2 and DVDs, but can also provide quality video even at 512 kbps.

Most digital security cameras use MJPEG, MPEG-4, or the new MPEG-4 version 10 known as H.264. H.264 (MPEG-4 AVC/H.264) is used on Blu-ray discs, which have about twice as many pixels as the same size DVD movie disc but provide more detail. Unlike the typical DVD player and MPEG-2 movie, H.264 requires massive amounts of horsepower because there are more mathematical computation requirements necessary to achieve the higher quality image at the same 6 MB per second. H.264 is a new, very important breakthrough in digital video surveillance. H.264 can encode and decode video at a higher quality, using half the bandwidth of any of its codec predecessors. It can provide better quality video for video analytics (see Chapter 10) and use less storage space.

This revolutionary compression technology was developed as a joint effort with the ITU and MPEG. H.264 can transmit HDTV programs twice as efficiently as MPEG-2, store 2 hours of HD-quality movies (Blu-ray) on an ordinary red-laser DVD, facilitate the emergence of high-definition personal video recording for consumers, improve the quality and storage capacity of HDD camcorders, and deliver CIF-quality video to mobile devices (Table 2-5).

Table 2-5   MPEG Standards Table

Version	Common Application	Bandwidth (2 h)	Storage
MPEG-1	Digital Multimedia Video	1.8 Mbps	2 GB
MPEG-2	DVD	3 Mbps	2 GB
MPEG-3	MPG3 music	N/A	
MPEG-4	Video streaming and recording	1.75 Mbps	1.3 GB
MPEG-4 v10 (H.264)	Video streaming and recording	1.1 Mbps	700 MB

Multiple Streams and Archiving

Somewhere out there in the midst of the Internet there are servers that store all the video made available through Web sites such as YouTube and Fancast. It doesn't magically appear out of the ether, but is stored on hard drives inside Web and media servers. All digital media takes up hard drive space somewhere. The higher the resolution, the more pixels, and thus, the larger the storage requirement.

The typical DVD movie fills all 4 GB of DVD space. However, viewing that level of quality streaming through the Internet is cumbersome, erratic, and unpleasant or impossible, depending on your connection speed. If a digital security camera required 4 GB of space for 2 hours of footage, it would result in 48 GB of data per camera, per day, or about 1.4 terabytes (TB) a month. Obviously this isn't cost-effective, as the storage requirements would be huge and the camera alone would be too expensive.

CIF video resolution is the mainstream for analog cameras, as most products offer CIF resolution. CIF is widely used for two reasons: (1) the digital video stream monitoring requirements aren't too high, as video transmission bandwidth is limited and (2) hard drive space is limited. One day of archived video at CIF resolution with a 512-kbps bandwidth ceiling is about 4 GB of storage space per day, recording at 30 fps and 24/7 (check out both www.genetec.com and www.axis.com for storage calculators) – far less than the 48 GB mentioned previously.

Many encoders and IP cameras offer multiple streams. The reason for this is to provide options for various requirements. One stream can be the recording stream, set to 30-fps CIF resolution at 512 kbps, to secure the hard drive space needed. If the camera was set on a variable bit rate it would be next to impossible to determine the exact hard drive space requirements, as the bitstream can be anywhere from 512 kbps to 5 Mbps at any given time. A second stream can be the viewing stream, set to a higher resolution, and a third can be smaller for remote viewers (reaching in through the sporadic Internet).

How MPEG Compression Works

Next is an abbreviated example of how digital MPEG compression works and explanation of why digital encoding requires heavy-duty processor horsepower. More details can be found at www.mpeg.org.

MPEG video is a hierarchy of four layers. These layers include the data found in most digital files, helping with error handling and synchronization. The first layer is the video sequence layer, which includes a bitstream. The second layer includes the reference to a group of frames that make up the pictures in the bitstream. The third layer is the actual picture layer, and the last layer is called the slice layer. The slices include a sequence of raster-ordered macroblocks. These macroblocks consist of four 8 × 8 luminance (black and white) blocks and two 8 × 8 chrominance (color) blocks. Macroblocks are used for motion-compensated compression. The 8 × 8 blocks are used for discrete cosine transform (DCT) compression, which separates the signal into separate independently streamed bands. The metadata for each independent band is in its header, so the codec understands where and when this stream gets encoded and then decoded (see Figure 2-8).

Frames can be encoded into intra-frames (I-frames), forward predicted frames (P-frames), and bidirectional predicted frames (B-frames). The I-frame is encoded as a single image with no reference to any past or future frames, thus becoming a "key frame" – a reference point for future frames. The data for each frame and 8 × 8 block are quantized. Quantization is a complex process of ignoring less important bits of the image; for example, the muddy detail of the brick wall behind the shadow of a tree. Quantization and subsampling are the only two lossy steps of the whole compression process. The resulting data are then encoded in a zigzag order to optimize compression by producing longer runs of 0s with little high-frequency information. In other words, it looks for all the same low-frequency data in each of the 8 × 8 blocks and delivers them first as the bulk of the data.

A P-frame is encoded relative to the closest preceding I-frame (or key frame). Each macroblock in a P-frame can be encoded either as an I-macroblock or as a P-macroblock. The I-macroblock is encoded just like a macroblock in an I-frame. A P-macroblock is encoded as a 16 × 16 area of the preceding I-frame, based on movement in the frame. If there are multiple areas of motion, such as two different cars crossing an intersection, then more motion vectors are added in relation to those positions. Motion vectors consist of half-pixel values, in which case pixels are averaged.

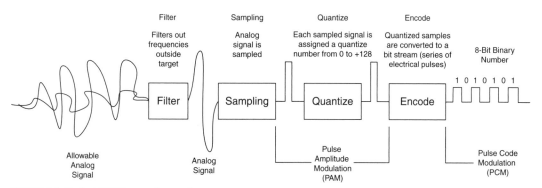

FIGURE 2-8 Typical MPEG encoding path.

An error variable is also encoded using the DCT, quantization, and run-length encoding. A macroblock may be skipped if there's no movement, which is equivalent to a (0, 0) vector, thus saving time and space for additional motion vectors and encoding. A B-frame uses the past reference frame, the future reference frame, or both. The future reference frame is the closest I-frame or P-frame. The only difference between a B- and a P-frame is the use of reference areas in a future frame, where the two 16×16 macroblock areas are then averaged.

Video frames, each one a single picture cell, are normally transmitted at a rate of 10-30 fps. Each frame is compressed and then dissected into smaller bits and bytes for transmission over IP networks. Frames may be compressed independently of other frames, such as I-frames or key frames, or compressed based on the differences from the P- or B-frames. These frames make up what's called a "group of pictures" (GoP), composed of an I-frame and variable number of P- and B-frames, which only encode the motion vector changes from the previous I-frame. For example, a GoP may include the following frames:

I, P, B, P, P, B, B, I, B, B, P, B, B, P, B, B

These GoPs are sent at 30 fps. The I-frame is independently compressed, resulting in a larger number of IP packets than the P- or B-frames, which only encode changes from the previous frame. These consistent variations to produce the best, most exact motion vectors, along with the smallest error rate and the best compression, the fact that MPEG-4 v10 (or Blu-ray's H.264) breaks down the macroblocks to 4×4 (in addition to 8×8 and 16×16), creation of a new flexible macroblock ordering (FMO), and slices, are all why the new H.264 video encoders and cameras demand enormous processing power.

Figure 2-9 shows three hypothetical frames from an MPEG-4 video stream. In this example, the I-frame sets the scene with a single key frame. Thereafter, in the relentless pursuit of shrinking bandwidth, the P- and B-frames, within their macroblocks, change that initial key frame whenever there's movement – in this case, the vehicles in traffic. The only additional bandwidth required is for the small macroblocks of movement in the upper left-hand corner (Table 2-6).

Fixed digital IP cameras are only concerned with movement within the area of coverage, but when you also add panning, tilting, and zooming to the equation, many PTZ IP cameras have a difficult time processing a consistent 30 fps. This, of course, is only a minor stumbling block as newer cameras include faster multiprocessor chips.

Analog Camera and Digital Video Encoder Versus the IP Camera

When any digital network camera is set to CIF, as is any camera with a digital sensor chip, it encodes just the 352×240 pixels it needs (or 84,480 pixels per frame). However, due to the MPEG-4 algorithm, there are only 84,480 pixels per *key frame* (every fourth

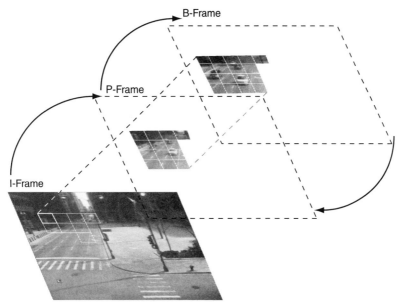

FIGURE 2-9 The I-frame presents a whole frame, while the P-frame and B-frame change the movement within the I-frame (here only the traffic on the street).

Table 2-6   Added Features for H.264

Feature	MPEG-2	MPEG-4	MPEG-4 H.264
I-, P-, and B-Frames	Yes	Yes	Yes
Interlace	Yes	Yes	Yes
Coding	Huffman Algorithm	Huffman Algorithm	Huffman Algorithm
Macroblock Size	Fixed 16 × 16	Fixed 16 × 16	Variable down to 4 × 4
One-Quarter Pixel Variant		Yes	Yes
Deblocking Filter			Yes
Multiple Reference Frames			Yes
Slice-Based Motion Prediction			Yes
MB AFF (Improved Interlaced Management)			Yes

frame) and never all 30 fps to minimize bandwidth. An analog camera connected into a digital video encoder, on the other hand, has a consistent 29.9 frames of 378,000 pixels to work with when encoding to MPEG-4. The MJPEG stream on a digital IP camera looks better because MJPEG forces full frames for each frame per second.

The analog camera/digital video encoder combination can easily grab 2,534,400 pixels (or the 1,500,000 pixels requirement) per second from the 11,302,200 pixels from the analog signal. Thus, an analog camera/digital video encoder combination can more easily reach 30 fps on CIF using MPEG-4 while the IP camera struggles, even with a heavy-duty processor.

Chapter Lessons

A few things to remember about digital video and media:

- Bandwidth and hard drive space are crucial components in determining the level of digital video quality required.
- All digital media presented on any screen (e.g., computer monitor, plasma TV, or LCD HDTV) are typically 72 or 96 dpi, while the same level of quality for print is typically around 300 dpi. This makes what appears on screen about a quarter of the size on paper.
- The traditional CCTV NTSC or PAL resolutions are becoming a thing of the past and megapixel cameras will become the premiere solution for DVS, thanks largely to MPEG-4 v10 or H.264.
- The pixels of any digital image (static or video) are the foundation of any and all generated, streamed, stored, or displayed digital media.

3

Digital Video Hardware

The Evolution of Video Surveillance Hardware

Video surveillance hardware has followed the same evolutionary track as photographic cameras. The days of a mechanical camera, when a single-lens reflex (SLR) used a cadmium sulfide photocell as a light meter (light sensor) to assist in the correct aperture and shutter speed configuration, are gone and the world is now run by the microprocessor. This is even more true with anything digital. This higher level of sophistication now requires a three-dimensional maintenance and troubleshooting view: mechanical, electrical, and software. The only power over a malfunctioning device is via software by upgrading the firmware (embedded software that runs the hardware), and if there isn't an updated firmware version to fix problems, then the only options are to either live with the problem or return the device.

Although many video surveillance cameras appear to be similar and provide the same specifications, they aren't the same. Don't assume the capabilities of any hardware from the marketing brochure. Digital video is still relatively new and advances are occurring in leaps and bounds, but as new technology is developed it must be tested and there's nothing more frustrating than finding out you're an unintended beta site for a new piece of hardware.

How Cameras Work

A video security camera isn't much different from an ordinary camcorder, although digital video security (DVS) cameras are designed for lower lighting and have longer life expectancies. Advances in video security cameras parallel motion picture technology from using film to videotape, and now to digital video files for archiving. It doesn't matter how technologically advanced cameras become, because they must all still follow the laws of optics: they still require a lens, focus, aperture, and shutter speed settings and must follow all the optical laws of depth of field. The lenses have essentially stayed the same although they're now computer designed for better optics at a lower cost. Although many of the lower cost security cameras use plastic lenses, glass optics still provide the best results.

Refraction

Light refraction is the changing path of light rays as they pass through select objects, which in our case is the lens element. A lens element is merely the physical piece of glass used inside a photographic lens. The more the lens needs to do, the more lens elements are required (zoom lenses have more lens elements than fixed lenses). The more lens elements, the more light refraction occurs, slowing the speed of the light and limiting the quality of the image.

Optics

In the traditional world of film cameras, shutter speed, f-stop aperture, and film speed were essential components to shooting static or motion pictures. There's a check and balance system when choosing the right configuration for the right type of photograph. In the days of mechanical cameras this was done as a slow and deliberate process. Once electronics were added to the mechanical machines some of the mystery was erased, but there were sacrifices made because the camera didn't understand the subject. Obviously, electronic cameras have come a long way, but cameras are just electronic machines that don't understand the subject matter they're required to capture.

A friend of mine asked me, because of my extensive experience in photography, why his high-end Canon digital SLR was taking blurry pictures. They weren't blurry pictures, but rather people and objects in motion. Any electronic camera, set to fully automatic, will choose what it perceives as the best configuration for that photograph based on lighting, the limitation of the assigned ISO setting, and the available shutter speeds coupled with the widest aperture. To capture motion or freeze frame the image, the fastest shutter speeds are required (see the section Shutter Speeds).

Although the photographic camera has been around since the nineteenth century, it wasn't until 1960 when the first light meter, a selenium photocell, was placed behind the lens for better photographic accuracy. This is an important advance, because it's the light that travels through the lens that creates the imagery. That was true in 1960 and it's still true today. The selenium photocell may have been replaced with charge coupled device (CCD) and complementary metal oxide semiconductor (CMOS) sensors (see the section Digital Image Sensors: CCD Versus CMOS), but the camera hasn't changed much at all.

F-Stop

Most lenses used in any photographic application are measured in millimeters and f-stops (pronounced "EF-STOPS"). The f-stop is the method of measurement for the iris opening of the lens aperture. All optical lenses, from high-end motion picture cameras used in Hollywood to the low-cost digital still camera in your pocket, use f-stop as a method for measuring the aperture of the lens (shown in Figure 3-1).

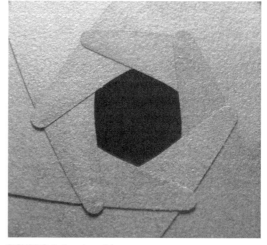

FIGURE 3-1 Aperture iris.

Lenses are labeled with their widest f-stop, which is the largest iris aperture capable by the lens. Historically this has been important, as the larger the lens aperture the more light passes through the lens elements and hits the sensor. More light means better imagery and lower light capability. An optical lens works like the human eye: when it's dark, the iris of the eye opens wider to allow in more light and when it's brighter, the iris contracts. However, since the optic nerve and the human brain are far more sophisticated than the typical camera sensor, the camera (still or motion picture) needs to work harder to accomplish what humans can achieve naturally.

F-Stop, Shutter Speed, and Depth of Field

Figure 3-2 shows a photo (labeled f/3.5) that was taken using a 55 mm f/3.5 Nikkor lens, set at f/3.5 and 1/1000 shutter speed. The f/3.5 lens aperture setting is the largest opening, allowing the maximum amount of light to pass through the lens to the sensor. This amount of light provided the ability to use a 1/1000 of a second shutter speed.

The other photo (labeled f/22) was taken using the same lens, but with the lens aperture set at f/22, the smallest lens opening. This shifted the burden to the shutter speed to make up the loss of lighting by dropping the shutter speed down to only 1/10 of a second. Fortunately, I was using a tripod, as handheld photography must be above 1/30 of a second; otherwise the image will blur from handheld movement.

Notice the difference between the two photos? In the first photo, at f/3.5, the background is completely blurred, while the second photo at f/22 provides an image that's clear in both the foreground and background. Thus, the smaller the aperture, the longer the depth of field.

Shutter Speed

The speed of the shutter controls the amount of light that touches the film and/or sensor. The slower the shutter speed, the more time for light to penetrate and illuminate the image on the sensor. The faster the shutter speed, the less light available to create the image. Shutter speed also controls how well the camera captures movement. Faster shutter speeds are able to capture

FIGURE 3-2 Depth of field.

movement clearly, whereas slower shutter speeds allow for more light but sacrifice clarity of moving objects. Figure 3-3 shows two photographs taken with a Nikkor 24 mm f/2.0 lens. The first photograph was shot at 1/10,000 of a second at f/2.0. The moving arm was captured in motion thanks to the faster shutter speed, but the blurred background lacks depth of field. The second photo was taken at 1/30 of a second at f/22. The slower shutter speed blurs the motion of the arm, but the higher f-stop (smaller iris aperture) lengthens the depth of field to include the entire background.

The advantage that video (security) cameras have over still cameras is the ability to create up to 30 frames per second (fps) or 30 pictures per second. This not only creates the illusion of motion, but also allows slow shutter speeds to increase the amount of light traveling through the lens for low-light environments. This feature, along with the use of lenses with faster (wider iris) f-stops, is how some security cameras can literally see in the dark.

FIGURE 3-3 Shutter speed.

Millimeters

The "focal length" of any lens (as of 1950) is measured in millimeters. The focal length of any sophisticated lens is complex. If the lens has one glass lens element (as many throwaway pocket cameras do) and is designed to be set to focus on infinity, then the focal length is measured from the center of the lens element to the focal point. However, sophisticated camera lenses include multiple lens elements set at varying distances apart. In these cases, the focal length is considered the distance from the secondary principal point (or secondary "nodal" point) to the rear focal point of the lens. Telephoto lenses have the secondary principal point in front of the first element of the lens, while wide-angle lenses tend to have their secondary principal point between the last element of the lens (closest to the sensor) and the focus.

In traditional 35 mm film photography (both still and movie cameras) the sensor fills the entire 35 mm frame. The size of the sensor determines how many millimeters the "normal" lens would be: that is, the lens size that shows the imagery at about the same size as the human eye. The full 35 mm sensor produces enough detail to use the 50 mm focal length as the traditional "normal" lens. That isn't the case with digital still, movie, or security cameras as they use both a smaller size and different type of sensor.

Digital Image Sensors: CCD Versus CMOS

The image sensor of the camera is responsible for converting the light and color spectrum into electrical signals for the camera to convert into zeroes and ones. All commercially available digital cameras (still, movie, or security) use one of two possible technologies for the camera's image sensor: CCD or CMOS. CCD sensor technology is specifically developed for the camera industry, whereas CMOS sensors are based on the same technology used in many electronic devices for memory and/or firmware.

CCD sensors have been used in photographic equipment for 20 years and up until recently included better light sensitivity than CMOS sensors. This higher light sensitivity translates into better low-light images, which is important for security cameras. A CCD sensor is more expensive to manufacture and incorporate into a camera than a CMOS chip. Thus, CMOS technology had to improve dramatically to meet the demand for lower cost digital image products. CMOS sensors are more cost-effective to manufacture and assemble, making smaller cameras with larger sensors possible, but they still lack the power of low-light sensitivity.

CMOS and CCD sensors are typically measured in either millimeters or inches. The majority of security cameras use anywhere from a 1/4″ to a 2/3″ sensor, which as you can see from Figure 3-4, is a fraction of the size of the traditional 35 mm sensor. However, that's why the "normal" lens on any digital camera is smaller than 50 mm. Even digital SLR cameras, although capable of using many of their 35 mm counterpart lenses, are considered 1.5–1.6 times their original 35 mm size (the 1.5× or 1.6× crop factor) because they're using the APS-C-sized sensor, which is smaller than the original 35 mm sensor.

Camcorders and security cameras may use smaller sensors, but in this case, size doesn't matter. What does matter is choosing the right lens for the sensor. Table 3-1 shows the "normal" lens (equal to the field of view of the human eye) for each sensor size.

Although the normal lenses for 1/4″ to 1/2″ sensors range from 6 to 12 mm, the function of most security cameras is to monitor a specific area of coverage and the 4 mm wide-angle lens tends to be the most popular (if not using a zoom lens).

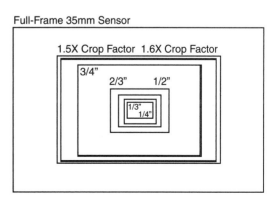

FIGURE 3-4 Sensor sizes.

Table 3-1 Normal Lenses and Their Sensor Sizes

Sensor Size (in.)	Normal Lens (mm)
1	25
2/3	16
1/2	12
1/3	8
1/4	6

Zoom Lenses

Zoom lenses are designed with multiple lens groups that when stretched or pulled together will change the field of view from wide-angle to normal to telephoto. The advantage of a zoom lens is the ability to grab closer visuals of a subject, but the disadvantage is limited f-stop aperture. Figure 3-5 shows an example of images a zoom lens can provide.

Fixed lenses tend to have larger iris apertures, offering more light to travel through the lens to the sensor for use in low-light situations. Depending on the zoom lens, the camera may also have additional limitations once set to telephoto. The surveillance requirements at the Navy Pier Phase I Project included installation of a professional high-end pan-tilt-zoom (PTZ) IP camera. This unit provides an angle of coverage using a 3.4–119 mm zoom lens. At 3.4 mm, its wide-angle position, the lens provides an f-stop of f/1.4, allowing the maximum amount of light through the lens and provides an area of coverage of about 50 degrees. At its 119 mm telephoto position, the lens aperture shrinks to f/4.2, reducing the amount of light and also narrowing the depth of field to only 1.7 degrees.

This 35× optical zoom factor is impressive enough, but the camera also provides additional 12× digital zoom capabilities whereas once the maximum optical zoom is reached, the camera will enlarge the image to simulate another 12 times its size. The quality of digital zooms is limited to the number of pixels available to upsample. The fewer pixels (e.g., CIF resolution), the fewer data the camera has to work with and

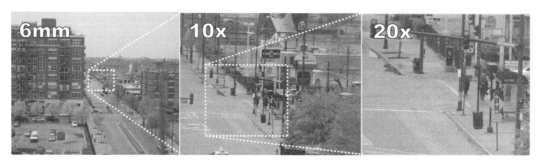

FIGURE 3-5 Zooming.

the lower the zoom quality. Digital zooms are far more impressive with higher megapixel cameras (as long as they're set to the maximum resolution).

Manual and Autofocus

Although the world has been overrun by autofocus cameras and camcorders, there are a few fixed cameras that require setting a manual focus, which considering the camera will never move, is a one-time event. Other than that, autofocus is recommended, especially for PTZ cameras. Autofocus provides cameras with the ability to quickly catch up with any automated or manual panning, tilting, or zooming. Fixed cameras typically don't offer autofocus functionality, only auto-iris, to control the amount of light captured day or night. Fixed cameras are installed, set up, and configured and since they never move from the set position, autofocus is unnecessary. This is detrimental if the fixed camera is a megapixel with digital PTZ, because if you pan or zoom into something that's out of the field of focus it will still be out of focus.

The best course of action when installing and setting up a fixed megapixel camera is to consider the specific requirements and determine what depth of field would be required. As mentioned earlier, the larger the aperture (opening in the iris), the more light enters the sensor, and the smaller the depth of field. The smaller the aperture, the greater the depth of field sacrificing the low-light capabilities. Multiple fixed cameras could provide some balance, if within the budget, as would moving to a professional PTZ camera.

Choosing the Right Cameras for the Right Job

As described in Chapters 1 and 2, there are two basic types of video surveillance cameras available today. The traditional CCTV analog camera, which delivers an NTSC signal in various formats and has up to 702 × 480 resolution (called 4CIF), and digital megapixel cameras, which begin at the higher resolution of 1280 × 1024 or 1.3 megapixels. These two types of cameras can also be IP cameras, which include a built-in digital video encoder to allow for immediate delivery of data over Ethernet. When using the same Ethernet cable they can be powered by the same RJ45 port. IP cameras come in various shapes and sizes, including smaller fixed cameras with vandal-resistant enclosures to full-blown PTZ cameras requiring separate power if used in an outdoor enclosure with a heating and cooling system.

The correct hardware to choose is based entirely on the requirements. In Chapter 6, Site Survey, I discuss various steps to help make those assessments if they haven't already been made clear by the customer. The details of the statement of work will determine how many cameras and what type will be of most value.

There's no real classification of security cameras and quality isn't necessarily coupled with a higher cost. Table 3-2 provides a summary of available cameras in three select groups based on cost and available features. For the purposes of this table, an economical security camera is considered anything that costs less than $200. The professional

Table 3-2 Available Camera Choices

Group	Economical (<$200)	Professional ($250-2500)	Specialty (>$20,000)
Fixed	X	x	x
Low-lux		x	x
LED IR	X	x	x
Thermal-IR			x
PTZ		x	x
Megapixel		x	x
Autotracking		x	x
Motion detection	x	x	x

products average about $2500 and specialty cameras are more than $20,000 and, depending on requirements, can be as much as $250,000.

Economical cameras are mainly products sold over the counter at various mass merchandisers, hardware, electronic, and camera stores. Professional security cameras include the name brand PTZ cameras, both analog and IP cameras, and can include ruggedized versions and enclosures, zero or low lux (for nighttime viewing), super zoom lenses, and multiple streaming and compression options. Specialty cameras that can provide human recognition up to a mile in pitch blackness or 2 miles during the day include a stabilizer and PTZ. Some of these specialty cameras are even more powerful with origins in military applications and the power to see for miles.

Economical Video Surveillance

There has been a growing market for economical security cameras, both wired and wireless. Companies such as Swann, D-Link, Linksys, and Lorex have developed some impressive fixed lens units for under $200 and even under $100. However, these cameras aren't professional video surveillance cameras. What makes a "professional" video surveillance camera? The same components that separate the pocket camera from the SLR or the economical camcorder from a handheld camcorder aspiring filmmakers use to make movies. The list includes, but isn't limited to, the following:

1. Lens quality
2. CCD or CMOS size and quality
3. Durability
4. Features
5. Longevity

Any professional photographer or filmmaker will tell you that what "makes" the camera isn't the camera itself, but the lens in front of it. The image that's recognized, converted, and saved comes through the lens and the level of quality of the lens is directly connected to the quality of the imagery. Security cameras that cost less than $80 use mass-produced single element lenses; some are even made of glass (grade B) and others are made of engineering-grade injection-molded plastic. The new wave of plastic lenses cuts the cost of lens manufacturing in half, thus opening up the market for lower cost cameras and security cameras. These cameras may work well indoors or in milder climates, but they aren't durable in severe weather climates – most clear plastic, even engineering-grade injection-molded plastic, deteriorates, fades, and fogs in harsh environments. If you're looking for a low-cost, short-term outdoor solution, these inexpensive cameras can deliver, especially with built-in LED infrared (IR) capabilities, but ultimately they're best used indoors. The economical cameras around $200 tend to have better quality lenses and components with highly respected brand names to support them.

Professional Video Surveillance

Professional camera solutions are built to last with high-quality Japanese- or German-made optics, power zoom lenses, higher-grade sensors, and outdoor enclosures (sold separately) made to withstand many years in severe weather conditions.

Figure 3-6 presents two images of the same egress. One is priced at $90 with 36 IR LED lights for night vision (to 30 feet) and is a fixed camera. This fixed camera not only uses a plastic lens, but is also less than 2 years old. Obviously, it didn't like the cold Chicago winters and hot summers. The other is a professional-quality PTZ dome camera priced at about $2800. Used versions are now available for a few hundred dollars and the one in the picture is about 10 years old, but will likely last for another decade. I'd like to reiterate that this is a printed representation of a digital image originally illuminated by the back light of a monitor. On a digital screen it may appear different.

Professional security cameras are also designed with vandal-resistant enclosures with unique security screws, making them difficult to access (see Figure 3-7).

Specialty Video Surveillance

The specialty category covers some of the more unusual requirements, such as human recognition at 1500 m (1 mile) at night; professional stabilizers above and beyond what's available in most professional cameras; and ultra-powerful, high-performance zoom lenses providing human recognition at up to 9000 m (6 miles).

FIGURE 3-6 Plastic versus glass lenses.

Long-range camera models include the Vumii Discover Series, Quickset GeminEye, Eaglevision EV3000, the Sierra Pacific IR360, and FLIR (which provides the thermal imaging technology for most manufacturers). Vumii is one of the global leaders in the advanced night vision video surveillance market. The Vumii Discover series provides visibility from as much as 6 miles away during the day and up to 3 miles at night (with a clear line of sight), using a continuous-wave laser IR illuminator that synchronizes with its CCD sensor.

The difference between these cameras and the economical IR cameras is their ability to "see" for hundreds and

FIGURE 3-7 Vandal-resistant enclosures.

even thousands of meters during both day and night. Like most video security cameras, long-range camera models are available as fixed cameras with fixed lenses or PTZ cameras with powerful, long-range zoom lenses. Obviously, cameras this powerful aren't sold at the local retailer and they're very costly. Thermal cameras are expensive because of the cost of the special germanium glass composite used in the lens and enclosure to keep it invisible to the thermal-IR sensor. The camera must read the IR energy of any object generated at the atomic level, but it must do so beyond the lens glass and enclosure.

Analog Versus Digital Versus Megapixel

When comparing security cameras, it's important to take note that each type of camera provides a unique function. As the convergence of digital technology continues to improve image quality on lower cost security cameras, manufacturers will eventually offer the best of each world.

As discussed in Chapter 2, the traditional CCTV NTSC analog camera has a foundation measured in television lines. The resolutions aren't much more than what an original analog television can display, even with a digital video encoder, as the encoder only digitizes what the camera delivers. Digital IP cameras, with their built-in encoder, can also be NTSC cameras manufactured for a CCTV environment, or they can move into the next generation of DVS security cameras and offer megapixels.

Megapixel cameras are different because they break the mold of this traditional CIF and 4CIF analog video image of 84,400 and 334,000 pixels, respectively. The megapixel camera begins at about 1 million pixels (1.3 MP) and increases from there. As discussed in Chapter 2, the more pixels used, the more detail provided, resulting in a higher quality digital image. Figure 3-8 shows the different images produced using low- and high-end PTZ as well as megapixel cameras.

NTSC IP cameras appear to generate better image quality than their analog camera/digital video encoder counterparts, but they do have two primary weaknesses: frame rate and low-light capabilities (limitations of the microprocessors and CMOS sensor). IP cameras generate an image directly from digital data. There's no continuous stream of analog video delivered so the microprocessor inside an IP camera works more than in the analog camera/digital video encoder pairing. This heavy processing usage makes it difficult for many IP cameras to generate a continuous stream of 30 fps at lower resolutions, especially PTZ IP cameras that try to generate 30 images per second and also need to compensate for panning, tilting, zooming, and computation-heavy codec such as MPEG-4 and H.264.

Swann Fixed Camera

Pelco Spectra PTZ Camera

Arecont 3MP Fixed Camera (with Digital PTZ)

FIGURE 3-8 Low-end, high-end PTZ and megapixel cameras.

There are many benefits to the IP camera, including the simple integration onto existing network infrastructures in multiple locations and even internationally, as Ethernet is a worldwide standard. The IP camera can also be physically smaller than analog cameras and doesn't need proprietary hardware to function (like matrix switches or controllers). IP cameras offer an optional progressive scan or deinterlacing (vs. interlaced scanning), which generates better video and static images, avoiding the interlaced shutter blind artifact distortion. Most IP cameras also have a "local recording" option to provide users in the Web interface the ability to record video locally on their own computer.

There is, however, a lack of standardized video protocol for all IP cameras, which greatly limits their interoperability with select Video Management System (VMS) software interfaces. Before choosing an IP camera for a solution, research and determine if that camera is compatible with the select VMS solution. Unfortunately, your research may need to include more than just reading the datasheet, which may indicate standard "MPEG-4" or "MJPEG," but manufacturers reach that point using different methodology and it may not be recognized by all VMS software.

30 fps

Newer PTZ IP cameras have improved by moving toward the new H.264 codec, although even that needs more horsepower. The problem isn't the camera's inability to achieve a continuous stream at 30 fps, but the overall concept that 30 fps isn't necessary because the human eye can only see 7 to 10 fps. However, the implementation of digital video surveillance isn't meant to limit the monitoring of the area of coverage to the human eye or to NTSC analog video. The technology must compensate for the limitations of human resources and 30 fps can be one of those advantages.

Let's set aside the fact that most casinos and military applications require 30 frames or more per second for video surveillance. One of the primary debates for using 30 fps is the amount of bandwidth and storage required to sustain it. Recording at 15 fps can cut the storage requirements in half, adding more room for existing applications or expansion. However, storage limitations for DVS aren't due to the number of VHS tapes and the manpower needed to swap out those tapes, but the amount of hard drive space, which is far more cost-effective. Bandwidth is only limited to the DVS network and the processing power and network adapter of the VMS server.

Recording 30 fps allows the DVS system to see beyond human capabilities, catching little nuances that may occur in a split second or even multiple angles for object recognition. If a human or automobile is moving, twisting and turning in different directions, the chance to capture a good facial photograph or angle of the license plate may be missed in a split second unless you have an option of 30 different frames to choose from within that second. Figure 3-9 is an example of a person looking around before moving out of camera shot. All 30 frames happen within a single second, with a few frames highlighted to provide more choices for recognition. Any camera shooting at 30 fps also includes faster shutter speeds and the ability to better capture motion (notice the slight blurring in Figure 3-10). Once the camera has zoomed in to the figure for recognition (as is the case here), the area of coverage is smaller and if the figure flees he would be in and out of the camera frame in a fraction of a second, making each of the 30 frames that much more important. See Figure 3-10 for a comparison at only 6 fps.

Power Requirements

Security cameras are typically powered by low voltage – from 12 V up to about 50 V over PoE. This is based solely on the camera specifications. Most economical security cameras use inexpensive 12 V power adapters, which require an 110 V power receptacle, whereas midrange cameras need 24 V. It's important to accurately distinguish which type of low voltage the camera uses, as 24 V DC (VDC) is different from 24 V AC (VAC). A 24 VDC power

FIGURE 3-9 30 fps.

supply or adapter won't power a 24 VAC requirement and vice versa unless the camera specifications determine that the power board inside the camera is capable of distinguishing between the two currents. Many cameras also provide an option of more than one type of power. Select IP cameras can be powered by 24 V or PoE, while others just provide an option of 24 VAC or 12 VDC.

Whatever the power specifications, remember that without power the camera can't perform, so don't skimp on the power supply or else plan on replacing it in a few years. Commercial power supply manufacturers such as Altronix and Cantek provide fused multichannel and even multi-voltage power distribution solutions designed for video surveillance products. See the section Troubleshooting Camera Power later in this chapter for more information.

The Remote Reboot

When a security camera is installed out of arm's reach, it's not easy to recycle power unless you have access to the dedicated circuit breaker. This is likely locked behind a door in an electrical room some-where or if on the street, inside locked traffic or lighting controllers with accessi-bility only by the proper authorities. Why the reboot? DVS includes many different components and modules, all of which contain electrical, mechanical, and soft-ware elements, whether that software is an operating system that hosts the VMS or the embedded firmware that runs each device behind the scenes. This is where a "three-dimensional trouble-shooting" methodology is necessary.

FIGURE 3-10 6 fps.

When dealing with any type of software (not just Windows), data can sometimes become corrupted or the communication link overflows between the device and the server and the only immediate solution is to break that repetitive cycle through a reboot. A remote power controller provides the administrator the ability to break that cycle through the network using a Web interface. The remote power controller is simply a Web interface that offers the ability to manage the power outlets on the device with power up sequencing and manual shutdown. Keep in mind that if connected via remote radio and/or switch, the connection will be lost upon shutdown (so choose restart rather than shutdown).

The Dataprobe iBoot is a cost-effective single outlet solution. Manufacturers such as Puluzzi or LPC offer multiport devices, both modular and rack-mounted. These devices can help when the video surveillance malfunctions under the following conditions:

1. The encoder and/or camera crashes or loses connectivity to the VMS application.
2. When using an analog camera and digital video encoder, this helps to determine if the camera or the encoder is faulty.
3. Some analog cameras require the digital encoder be active when communication is activated. This may require a timed delay for the camera to boot up.
4. Corrupted or "hanging" firmware or software.

Lighting Illumination

There are two types of light sources that affect photography: primary and secondary. Primary sources include the sun, moon, street lights, and computer monitors. Secondary light sources are objects that can reflect light, but can't generate it.

Luminous intensity is the illuminating power of the primary light source and is measured by units called candelas. A candela is approximately the amount of light of a single candle per square foot. Luminous flux is the luminous intensity at a select angle, obtained by dividing the foot-candles by 4π radians and is measured in lumens. The number of lumens per square meter is called lux. Lux is used to provide specifications for the low-light capabilities of a security camera. However, each manufacturer uses different configurations to achieve the best possible lux factor. For example, if a camera has the capability of 1/15 of a second shutter speed with a 6 mm f/1.4 lens, those two elements, coupled with low-light upsampling technology, can provide an impressive low lux factor. But that may not be the most popular, or the most effective, configuration for the required area of coverage.

Figure 3-11 shows the light distribution of a 1000-W high-pressure sodium lamp mounted at 30 feet without any degree of tilt (perfectly horizontal). The light distribution using a metal halide lamp is similar, but doesn't quite reach the same levels. These are the two most popular lighting systems used for outdoor commercial lighting. The chart shows that directly underneath the 1000-W bulb (which is 30 feet above), the light measures about 15 foot-candles (candela) per square foot, or the equivalent of 15 candles lit within 1 square foot. That illumination drops to about half a candle per square foot at 60 feet. The ability to increase the illumination of an area of coverage

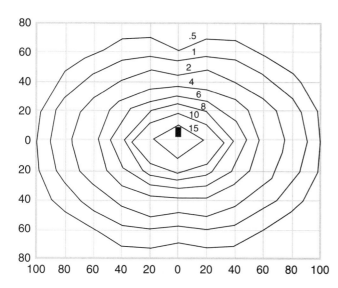

FIGURE 3-11 Measurement of illumination from a 1000-W bulb on a 20-ft. pole in foot-candles.

would require many additional lamps and power or low-light cameras to compensate for the lack of lighting.

Day and Night Cameras

The Pelco, Axis, Sony, Sanyo, and Panasonic are all professional PTZ IP cameras, each with up to 35× optical zooms and additional digital zoom capabilities. They all have similar area-of-coverage capabilities; typically from about 2 degrees at telephoto to about 50 degrees at the wide-angle setting with the standard zoom lens. There are a few differences between brand names, one of which is the low lighting factor.

Actual starlight, without a moon, is measured at 0.00005 lux. A moonless clear night sky (in an urban area, as rural areas tend to be much darker) is measured at 0.001 lux, and if overcast, that drops to 0.0001. The full moon, on a clear night, raises the lux factor to 0.25. Twilight is measured at 3 lux.

The Axis 233D camera in black and white can "see" down to a lux factor of 0.008, so without any additional lighting at all, the 233D can record recognizable imagery at twilight or with nothing but a full moon on a clear night. The Pelco Spectra IV has a lux factor of 0.00018, which means it can record and "see" imagery with nothing but a moonless clear night sky (see Figure 3-12). The Panasonic NW964, with its Super Dynamic III upsampling technology, rivals the low lux factor of the Pelco Spectra series, but is only specified as 0.013 lux. Although these three examples cost a few thousand dollars each, they provide a number of professional features and can record effectively at nighttime without the need of thermal-IR sensors, IR illuminators, or additional lighting.

IR technology has not only improved, but also has been added to cameras as a cost-effective solution by the more economical camera manufacturers. The inclusion of IR technology provides manufacturers with the ability to incorporate a simpler, cheaper lens and overall camera solution. It's significant to say that video security requires 24/7 day- and nighttime surveillance and the new economical solutions make it possible to

FIGURE 3-12 Day/night shots from an IP megapixel camera, an economical 48 LED IR camera, and a fixed, high-performance, FLIR thermal camera.

do so with some limitations. While these cameras do provide zero lux, giving them the ability to see in pitch blackness, these IR LEDs are typically limited to a distance of about 30 feet. The reason for this limitation is due to the way IR illuminators work.

Infrared and Thermal Technologies

There are two types of IR security cameras available, each with very different technologies. Laser IR illuminators and thermal imaging are both part of the IR spectrum, which is measured by the amount of energy in a light wave related to its wavelength. Shorter wavelengths have higher energy, and conversely, longer wavelengths have lower energy. IR light is emitted by an object at the atomic level, making it invisible to the naked eye.

IR light can be split into three categories:

- Near-infrared (near-IR) – Closest to visible light with wavelengths ranging from 0.7 to 1.3 µm, or 700 to 1300 billionths of a meter.
- Mid-infrared (mid-IR) – Has wavelengths ranging from 1.3 to 3 µm. Both near- and mid-IR are used by a variety of electronic devices, including remote controls.
- Thermal-IR – Occupying the largest part of the IR spectrum, thermal-IR has wavelengths ranging from 3 to over 30 µm.

The key difference between IR illumination and thermal-IR is that thermal-IR "sees" the energy emitted by an object instead of reflecting IR light off an object using laser or LED IR illuminators.

Figure 3-13 presents a comparison of three types of cameras focused on a rural environment. These security cameras would react differently within a highly illuminated urban environment, using the reflection of multiple street and building lightings to make the scene more visible. But what would happen in a blackout? When this occurs, the ability to directly "see" the energy, rather than rely on reflected light, becomes essential. The first camera in Figure 3-13 is a fixed 2-megapixel, higher quality camera, with one image shot during the day and the other in the middle of a moonless night. The daytime photo is of exceptional quality with lots of detail in the flora and even in the sky. The night shot lacks all detail but a subtle light in the sky toward the horizon and the bright street lighting.

FIGURE 3-13 The urban environment makes it easy to see in the dark. This is a high-performance professional PTZ security camera during the day and then night, zooming into the entrance of the far building.

Next, the indoor/outdoor fixed 48 LED IR analog NTSC camera shows a daytime shot depicting a cloud covering and a nighttime shot with the LED IR lights only able to reach the bushes 12 feet. away, and everything else beyond the LED IR 30-foot limitation. The fixed FLIR thermal analog NTSC camera doesn't require any IR illumination because it reads the existing IR spectrum at the atomic level. Not only are you able to see the suspicious figure about 50 feet away but also the sky, horizon, trees, bushes, and lights.

IR illuminators are very popular with security cameras from the economical models up to laser IR illuminators added to long-range cameras. The economical brands typically add a few dozen LED IR lights around the fixed camera lens to reflect IR off the monitored objects. The addition of these IR LED lights provides the fixed camera with the ability to literally see in the dark. There are dozens of brands that offer this feature in a variety of styles from bullet enclosures to vandal-resistant domes, all for less than $200. These cameras typically use the same CMOS or CCD sensors as other security cameras, even the same (plastic and/or glass) lenses, but the one problem I've come across with these cameras is the same problem with most outdoor cameras – dirt. In this case, the dirt doesn't necessarily block the view, but it does confuse the light sensor, which is what determines when to turn on the LED IR lights. It also reflects the IR light onto itself, creating a washed-out, low-contrast image during the day.

Professional solutions to this problem include a separate IR illuminator installed next to a higher quality camera, or a laser IR illuminator that focuses the IR light on a pinpoint far away. Depending on the manufacturer, power, and sophistication, these solutions can cost thousands of dollars. Supplementing a PTZ camera with illuminators can be costly, as only one illuminator can light up the entire area-of-coverage capabilities of a PTZ camera with $35\times$ zoom.

How Weather Can Affect Cameras

Throughout their years in the field, the number one issue security cameras have with the weather isn't rain, sleet, or snow, but the residue left behind on the dome and/or enclosure windows that can obstruct the camera's view. This can become such a maintenance nightmare that a forecast of rain or snow brings on a sense of dread, not for the slippery roads, but for the cameras' dirty domes. Many dome cameras (Bosch, Pelco, Axis, Sony, etc.) offer a "water-resistant" resin coating or even wipers to help in the dissipation of water, but these are options and not standard features. Weather must be considered during the site survey to determine if any extraordinary features must be added to the requirements to avoid visibility problems when the seasons change.

One of the best ways to determine if the proposed devices are capable of meeting requirements is to look at other installations where the weather is comparable. This may provide some insight on how to best proceed and with what accessories or add-ons. Unfortunately, technology moves so rapidly and improvements are implemented so quickly, that many times each installation may have a slightly, or dramatically, different version of the product being considered.

There are also unseen problems that arise while testing in the laboratory. For example, camera housings with outdoor specifications don't necessarily mean outdoors along the lakefront in Chicago in blistering cold, subzero February weather. Any outdoor housing for professional video surveillance cameras includes a fan for cooling in the summer and a small heater for heating the unit and theoretically, melting any ice that may form on the housing or internally from condensation. There's a considerable difference between testing devices in a laboratory and testing them out in the real world. The only way to truly avoid an implementation failure due to weather is to test the devices in the proposed environment before starting a full-blown installation. It's not unheard of for sales and marketing to "elaborate" on a function or feature that, unfortunately, fails to perform to specifications thereafter. It's always best to check the specifications on all the equipment, check the support Web sites, and if possible, ask an engineer at the company to elaborate on the testing procedures.

■ ■ ■ ━━

Temperature and Climate Control

There are cameras, encoders, and even network devices inside "hardened enclosures" or "cases." This equipment is specially designed to withstand higher and lower temperatures beyond ordinary specifications. Whenever installing electronics in high-temperature or subzero locations, it's best to consider these for better performance and longevity. However, electronics tend to perform better in colder rather than warmer weather, but dramatic changes in subzero weather can cause condensation, which creates moisture that can be immediately detrimental to electronics or create latent failure problems. Extreme heat can cause problems with electronic devices that run software, as any overheated computer will demonstrate by freezing or crashing. Best practice when implementing expensive equipment is to provide a cooling and heating system (inside and outside) to keep the hardware at a consistent temperature, thus extending the life of the equipment.

━━ ■ ■ ■

Rain, Water, and the Weather
Water has a tendency to find its way into electronics, given the opportunity. The trick is to not give it any opportunity. Personally, I haven't experienced water in an advertised "outdoor" housing (water freezing on the outside of the housing is a different story), but rain and water do tend to find their way into wireless antenna connectors and cables unless tightened properly and weatherized (more on this and how rain affects wireless signals in Chapter 5).

Many installers find it easier to just tighten the end as much as possible, or a locking mechanism, if included, may be enough to keep the elements out. Just because a standard or proprietary connector is advertised for outdoor use doesn't mean that it doesn't need to be properly weatherized. Sealing the connectors, even with electrical tape, will extend the life of the device and/or cables.

Outdoor/Indoor Cameras

As previously mentioned, the economical video surveillance solutions, like any electronic device, have longevity in milder climates and climate-controlled environments. It's when installed outdoors in harsher climates that they fail to perform consistently. Professional video surveillance cameras, whether analog or digital, offer ruggedized outdoor enclosures or housings specifically designed to protect the camera from the elements and the imagery generated by those cameras. These housings include heaters, not only to keep the equipment warm in subzero temperatures, but also to prevent the dome or window from fogging up in cool and damp weather.

The problem with the small, fixed LED IR cameras is the plastic used in front of the lens and in some cases, the actual lens. After a few years in the harsher climates, this plastic tends to fade and become milky. This makes it difficult for the CCD or CMOS sensor to successfully translate the imagery, and if there's a series of LED IR lights surrounding the lens, it can cause a halo-like affect that blurs the image day and night. Indoors, these devices have a far longer life span, but outdoors, or even in a garage without climate control, many of them lose their sharpness and clarity succumbing to the harsh environment.

Fixed Versus Pan-Tilt-Zoom

A "fixed" camera is a still camera; it's set to one position and is unable to move unless the lens is physically moved. A PTZ camera gives the operator the ability to move the camera lens using the VMS software. While a fixed camera can be more cost-effective, a PTZ camera expands the area of coverage beyond a single fixed point. Depending on how the camera is mounted, a PTZ camera can provide 360 degrees of coverage and more than 35× the zoom capabilities.

Figure 3-14 depicts a close-up of a figure using the zoom feature of a PTZ

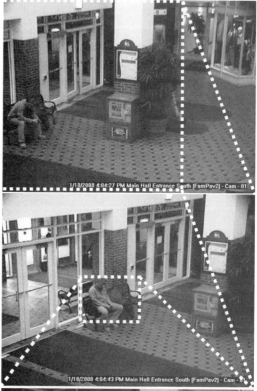

FIGURE 3-14 PTZ cameras, whether they're mechanical or digital, give you more flexibility in monitoring the area of coverage as you can follow a suspect or simply configure an optional autotrack function.

camera. The actual distance of the figure is shown in the first photo capture. The vast majority of today's PTZ cameras are analog NTSC cameras. These cameras, like the one used in this example, can be connected to digital video encoders (Verint 1500 series in this case) to convert the analog signal to digital, thus relinquishing the demand on processing power to keep up with the motion and movement to the encoder. PTZ cameras are valuable for DVS or for any video surveillance.

Eventually, megapixel cameras will offer comparable PTZ functionality, once the internal microprocessors reach the level of power required to generate a consistent digital image at megapixel resolution and continue to do so while the camera is panning, tilting, and zooming at the operator's request. Today megapixel cameras offer more pixels, thus more detail, and many offer a digital PTZ (i.e., the Arecont 2155DN) that can take the higher resolution imagery and digitally zoom into the pixels, upsampling and simulating a mechanical PTZ camera. Typically, these digital PTZ features work well up to a specific distance that stretches the resolution for a lower quality image. That distance is unique, depending on the number of megapixels. You can zoom into a 5-megapixel camera frame and get better imagery than a 2-megapixel camera because there are more pixels to work with.

■ ■ ■ ────────────────────────────────────

Telnet

Telnet is a terminal emulation program (client built-in to Microsoft Windows) for use on TCP/IP networks. The Telnet program runs on a computer and links the computer to a Telnet server on the network. Most IP cameras, wireless radios, and digital video encoders come with a built-in Telnet server, making it possible to access the device using terminal emulation. Telnet allows specific commands to be executed to the server software (and/or device firmware) as if entering them directly on the server console.

If the server and/or device includes a Telnet server function (TCP port 23 must be opened to allow Telnet access), the device can be accessed through the command prompt or through applications such as HyperTerminal or Putty.

──────────────────────────────────── ■ ■ ■

PTZ Protocols and Communications

The biggest problem with adding any analog PTZ camera to a DVS system is the communications required to convert the original PTZ communications between the camera's controllers to the new third-party digital video encoder. The analog NTSC signal generated by the camera is usually a standard signal that any encoder can read and convert to a digital signal. The PTZ controls are a separate connection and unless the camera includes a standard or at least a recognized protocol by both the digital video encoder and the VMS software, then an interface is required or there will be no way to take over the control of the PTZ function. Many older CCTV systems use a proprietary protocol that gives the main controller, its keyboard, and joystick the ability to control the PTZ function. This is done using hardware specifically designed to be used with such cameras and systems. DVS is about convergence and unless the proposed devices and their interfaces include a method of speaking to the camera's PTZ controller, then another option must be considered.

When you're dealing with digital conversion and are using a camera and encoder from separate manufacturers, they may have different methods of wiring the PTZ function. There are many standardized protocols for communications, usually using RS232 (limited to 50 feet), RS422, and RS485 (up to 4000 feet), each with a two- or four-wire option. The largest obstacle in determining how to correctly wire the camera to the new encoder is the support system, because camera and encoder combinations aren't usually included by the manufacturer. In other words, the camera manufacturer may point to the encoder manufacturer for an answer, while the encoder manufacturer may point to the camera manufacturer. This is where researching successful combinations, down to the exact model numbers of the camera and encoder, is important.

I can't tell you the frustration that mounts when attempting to match an analog camera to a digital encoder for PTZ control. When upgrading a CCTV system to a DVS system, or just adding a number of older, clearly high-end PTZ cameras to the DVS system, best practice is to determine if the camera is compatible with the new proposed system (both hardware and software) or if it needs to be replaced. This information is always good to know beforehand, but it's impossible to know how a select camera will react to a matched encoder unless tested in a laboratory. (Don't rely on the marketing materials or sales pitch!) If testing is out of the question, then make an assumption that those cameras are compatible and if not, they will need to be replaced at an additional cost. Be sure to document this on your site survey report; otherwise, without proper documentation you will likely be responsible for this cost.

Information about compatibility with the VMS software (at least the high-end professional solutions such as Genetec, OnSSI, and Milestone) is usually published by the software developer. Many lower end cameras and software solutions promote compatibility, such as "Pelco D compatibility," but lack the documentation to explain how compatibility was tested and with what encoder/software.

An IP camera can have similar issues, but the camera and encoder are one, such that both power and communications run through a single Ethernet cable. Just as it's important for an analog camera to communicate with the digital video encoder, the VMS software must have the appropriate drivers to control the camera, the encoder, and its functions. An IP camera has everything in a single unit, but it still must be compatible with the VMS system.

Recently, I had the privilege of working with some Bosch professional video surveillance equipment. Bosch's latest cameras and encoders provide a means of communicating with third-party equipment and software, but that's not the case with the previous Autodome models (originally manufactured by Phillips). The pre-Autodome G3 series isn't compatible with digital encoders without a converter interface device (believe me, I've tried). The Phillips/Bosch Autodome PTZ cameras were and still are one of the more durable analog video surveillance cameras on the market and a key focus of DVS is the integration of existing security assets. Bosch purchased Phillips (security cameras) and VCS (digital video encoders), although at the time of this consolidation these two systems weren't compatible. A Bosch VCS VIP 10 encoder can be configured to control the PTZ of a Pelco analog camera (using a proprietary VCS cable), but because the older Bosch/Phillips Autodome is part of an enclosed system using their own BiPhase protocol, they weren't compatible. Bosch recognizes this problem and has developed the LTC 8782 Code Translator, which can convert BiPhase to a number of different protocols and vice versa, if needed. There are also manufacturers such as Dante and StarCard that have developed PTZ control converters, so those existing assets can still be used. Keep in mind that depending on the VMS system software used, eventually a newer version may be developed and it will be time to upgrade to a system that provides open standards for better future growth.

Configuring PTZ communications with the analog camera can be done through proprietary software interface or, if available, through the encoder's Web interface. Manufacturers typically follow the traditional TX+, TX−, RX+, RX−, D+, and D− labeling, but there's no consistency in how to wire an analog PTZ camera to a digital video encoder as each camera and encoder is different. It can be a two-wire RS232, RS422, or RS485 or a four-wire configuration. You may find that many times, without the right documentation, this process becomes a matter of trial and error and the camera manufacturer's support center staff may recommend calling the encoder's support center staff and vice versa.

Along with the wiring variations the encoder can provide multiple choices in protocols. These options can be chosen from an "interface" or "serial" menu in the Web or Telnet interface of the digital video encoder (see Figure 3-15).

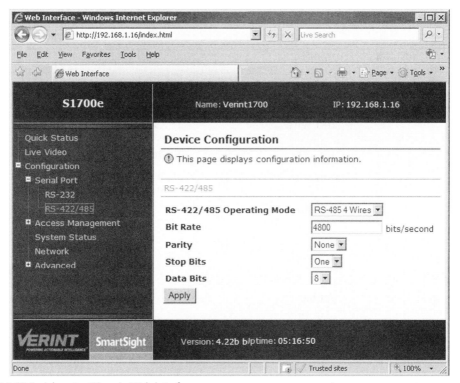

FIGURE 3-15 Serial port settings in Web interface.

■ ■ ■

Speed Limit

When choosing the PTZ communication protocol, it's important to choose the right communication speed. Setting the serial communications to 9600 Baud (bps) when the encoder can only understand data traveling at 2400 bps is another common mistake when attempting to set up the unit (see Figure 3-16).

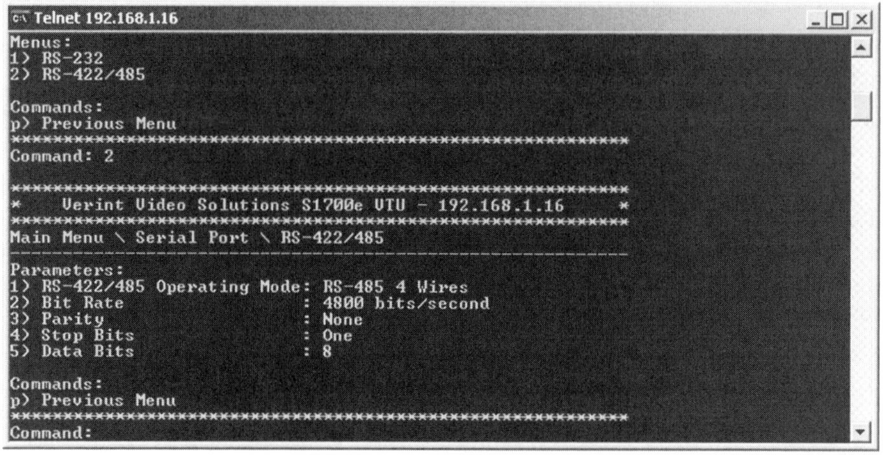

```
Telnet 192.168.1.16                                                    _ □ X
Menus:
1> RS-232
2> RS-422/485

Commands:
p> Previous Menu
*******************************************************************
Command: 2

*******************************************************************
*        Verint Video Solutions S1700e VTU - 192.168.1.16         *
*******************************************************************
Main Menu \ Serial Port \ RS-422/485
-----------------------------------------------------------------
Parameters:
1> RS-422/485 Operating Mode: RS-485 4 Wires
2> Bit Rate                  : 4800 bits/second
3> Parity                    : None
4> Stop Bits                 : One
5> Data Bits                 : 8

Commands:
p> Previous Menu
*******************************************************************
Command:
```

FIGURE 3-16 Telnet serial settings.

■ ■ ■

Digital video encoders come in many shapes and sizes. A single input port option will take a single analog camera and digitize the NTSC video for network connectivity. A digital video encoder simply serves the digital video to the VMS software, whether it's the software included on the CD-ROM with the unit or through a larger enterprise system. Verint, Axis, Bosch, and others also manufacture digital video encoders with multiple input ports, providing 4, 8, 12, or 24 input ports on one device with terminal blocks for PTZ control. The input option is typically the BNC coax connector, but there are a few older units that still require an RCA composite connector. However, there are many adapters available (even at the local Radio Shack) that can convert RCA composite connectors to BNC and vice versa.

Ruggedized for Harsher Environments

Like video surveillance cameras, digital video encoders are also available in ruggedized versions designed to withstand harsher environments.

Two-Way Audio

Many video surveillance cameras, in any category, provide audio capabilities. This function can be one-way where in addition to the video imagery the device offers accompanying audio surveillance. It can also be two-way audio, which provides the ability to monitor the video and audio and a means of providing audio direction through the camera or its accessory speaker. Audio also adds another layer of necessary compatibility with the select VMS software, so make sure the audio as well as the video is supported.

Configuring Digital Video Encoders and IP Cameras

The vast majority of digital video encoders can be configured using a DB9 RS232 serial port, using their own software, or Telnet using a default IP address assigned to the unit at the factory (see Figure 3-17). The RJ45 Ethernet port can only use Telnet and/or a Web interface (if they include a built-in Web server) if the unit was assigned the default IP address at the factory. The RS232 serial port also opens up the unit for the manufacturer's own configuration tool, using another protocol other than TCP/IP to communicate with the device. Many of the digital video encoder manufacturers have their own software application specifically designed to communicate with their product for the initial installation process and for troubleshooting and more advanced configuration.

Unit Discovery

Unit discovery is simply the process of adding a new digital video encoder onto the DVS network. If the unit is set with a factory default IP address this can be a simple process accomplished by using a Cat5 patch cable (or crossover cable, depending on the requirements). If not, then typically the manufacturer includes the installation software required to configure the unit to your specifications to add to the network.

Installation and Configuration Applications

The location of existing CCTV analog cameras may deter integration as existing assets because the power or data runs may be too far or traveling in the wrong direction from the assigned IDF. In that case, create an "environmental box" ("E-Box"), which is simply a Type 4 outdoor-rated NEMA enclosure with the single encoder and/or wireless radio inside to transmit the data long distances without having to run long runs of cabling and conduit (see Figure 3-18).

Most digital video encoders and IP cameras, like any computer electronics,

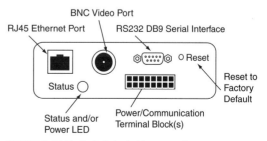

FIGURE 3-17 Typical digital video encoder.

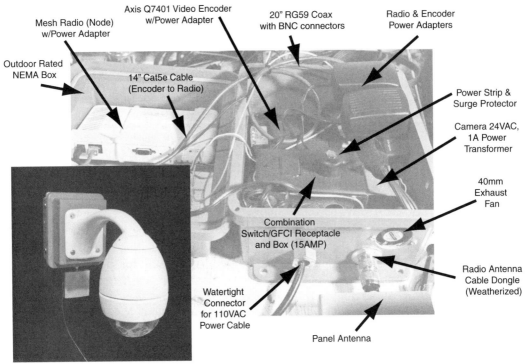

FIGURE 3-18 Outdoor enclosure with digital video encoder (to convert the CCTV analog signal into digital), a 24 VAC power transformer for the CCTV camera, a surge protector, and wireless radio to transmit the data wirelessly at great distances without the need of new cabling and conduit.

come with a CD-ROM for configuration. The CD may include the tools required for installation and a soft copy (PDF) of both the installation procedures and the operation manual. These application tools make it possible to communicate with the digital video encoder or IP camera rather than using Ethernet and TCP/IP. They may be configured using the RS232 serial port, USB port, or the RJ45 Ethernet port located on the front of the unit (see Figure 3-17). The device will include the power connector or terminal block, video BNC connector and/or connectors (if multiple ports), and optional audio connections.

To configure the digital video encoder or IP camera using the provided application tool, connect the digital video encoder or IP camera to a configuration laptop and insert the CD-ROM (or download the application from the manufacturer's Web site). Save a copy of the application tool's EXE file and install it (if applicable). A few of the application tools may be small, simple tools with the single task of communication with the digital video encoder and/or IP camera, so installation may not be necessary. Power on the digital video encoder and/or IP camera and connect them to the configuration laptop. Double-click on the application tool shortcut (once installed) on your Windows desktop and configure the application to communicate with the devices. This may involve setting the correct serial port settings or IP address (as determined by the documentation provided by the manufacturer).

The following scenario uses the Verint SConfigurator.exe application tool as an example. This tool is specifically designed to locate all Verint-only digital video encoders on the network and/or through the RS232 serial console connection. To configure the encoder using the console option, follow these steps:

1. Launch SCONFIGURATOR (doesn't require installation).
2. Choose CONSOLE (see Figure 3-19).
3. Press CONNECT (make sure the RS232 settings are correct).
4. The Verint Main Menu appears (see Figure 3-20).
5. Choose NETWORK and the network menu appears.
6. Disable dynamic host configuration protocol (DHCP) and type in the new static IP address.
7. Type "P" and press ENTER to return the previous menus and choose REBOOT.

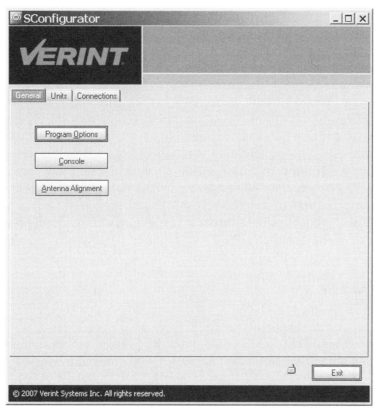

FIGURE 3-19 Verint's proprietary SConfigurator software is one way to discover a Verint device on the network and configure it. Most of the high-end digital video encoder manufacturers provide similar installation software.

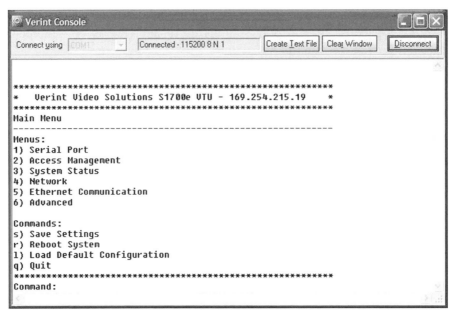

FIGURE 3-20 Verint SConfigurator console connection through RS232 (or you can use Windows' HyperTerminal).

DHCP

Many digital video encoders and IP cameras come from the factory with either an assigned IP address (which makes it easier to configure) or with DHCP enabled. DHCP provides configuration parameters to IP hosts. It consists of two components: a protocol for delivering host-specific configuration parameters from a DHCP server to a host and a mechanism for allocation of network addresses to dynamically configured hosts. Without a DHCP server configured to operate on the network, a DHCP-enabled device won't receive an IP address to join the network.

When the DHCP is enabled, the IP address range from the DHCP server will automatically supply the encoder with an IP address required to operate on the network. This can simplify the storage, tracking, and installation of the IP components by assigning the dynamic IP address using an allocated range of IP addresses and/or directly via the hardware-embedded MAC address. By grouping the dynamic IP addresses within a specific subnet mask, the encoder and any other device requiring an IP address to join the network can be more easily located and diagnosed.

DHCP isn't the best choice for security, however. If anyone with a laptop configured to receive a dynamic address plugs into any DHCP network, not only will that rogue laptop receive an address, but the user now knows the subnet details of the video surveillance network. The only way to avoid this potential security catastrophe is to provide the DHCP server with the MAC address of each device that has permission to be part of the network and turn off all other dynamically assigned addresses. Depending on the size and fluidity of the DVS network, this can be quite cumbersome. What happens if an encoder and/or switch fails and is replaced overnight? Without adding that MAC address to the allocated devices on the DHCP server, that device won't work and the maintenance personnel may be pulling out their hair trying to determine the cause for such an unusual failure of a perfectly good encoder.

Although it's more preparation and work, best practice involves manually allocating IP addresses for each device and not using a DHCP server. Also, the digital video encoder must hang on to the assigned IP address even after loss of power or an inadvertent reset to factory default; otherwise each time the unit loses power it reverts to its DHCP-enabled state. With no DHCP server to assign an IP address, the unit will need to be reconfigured every time. This is unacceptable, so always make sure the digital video encoder has flexibility with configurations and default adjustments.

Ethernet Configuration

To access any device on a TCP/IP Ethernet network, all equipment must be configured to be part of that network. This means each device must have that specific class of IP address and subnet. If not, there's no way to communicate using Ethernet. This can easily be accomplished with a laptop, one of the most important tools of the trade (more on this later). Once the factory default IP address is uncovered from the device documentation (or Web site), the next step in configuring the unit for the new DVS network is to have a laptop (or desktop) join the subnet network to which the device is already configured. First, follow this path to your computer's network connection icon:

Start ≫ Control Panel ≫ Network Connections ≫ Local Area Connection

Right-click on the local area network icon (see Figure 3-21) and choose PROPERTIES. Once the PROPERTIES dialog box appears (see Figure 3-22), highlight INTERNET PRO-TOCOL (TCP/IP) and then click PROPERTIES. Another dialog box appears providing two primary choices: (1) Obtain an IP Address Automatically or (2) Use the Following IP Address. If using a company laptop, it may be configured for DHCP, which is the first choice of obtaining the IP address automatically. If not, then an administrator may have provided a static IP address.

If the laptop uses a static IP address, make a note of the IP address, subnet mask, and gateway IP address (write it down somewhere accessible but secure). In its place, type in an IP address that matches the same network as the device in question, except not that same exact IP address. Remember, no two devices can share the same IP address. For example, if the default IP address for the new unit is 169.254.128.130, with a subnet mask of 255.255.0.0, then your laptop can use 169.254.128.131 with a subnet mask of 255.255.0.0. It can't have the same 169.254.128.130 IP address as the new unit. What happens if two devices on an Ethernet network have the same IP address? It depends on how the devices react to such an event. Microsoft Windows will issue a pop-up balloon that indicates a duplicate IP address error, but digital video encoders, switches,

FIGURE 3-21 Right-click on Local Area Connection.

and radios each react differently. Many devices just stop communications on the network until the conflict has been corrected; others will drop out of view until physically rebooted/power cycled.

Once you've entered the correct data, click the OK button and then CLOSE. Start any Web browser and enter the default IP address into the location bar, in this case http:// 169.254.128.130. If the unit is password protected then an ID and password prompt will require the default ID and password (typically "admin" or "root").

Commissioning Digital Video Encoders and IP Cameras

The commissioning process involves end-to-end testing of components previously installed, assuming the underlying support components are functioning properly. Once the digital video encoder and/or IP camera appear in the application tool, it's configured with the appropriately assigned IP address, subnet mask, and gateway.

Once an IP address is assigned, a list of specific configurations must be done within the Web interface (or Telnet, if a Web interface isn't available) prior to adding the camera/encoder into the VMS software. Those steps for the IP cameras (using an Axis camera as an example) are as follows:

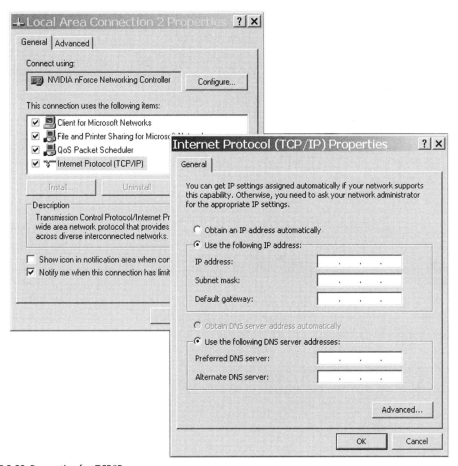

FIGURE 3-22 Properties for TCP/IP.

1. Under Basic Configuration
   a. Assign the new password to the User Permissions.
   b. Change the default Time settings to use the computer's time and date.
   c. Turn on the Daylight Savings Time radio button and choose Central Time: Chicago from the drop-down menu.
   d. Under Video and Image, make the following changes:
      i. Change resolution
         - 4CIF (702 × 480)
         - Compression at 30% (the lower the compression, the better the quality)
      ii. Choose the following Overlay Settings:
         - Check Include Date.
         - Check Include Time.

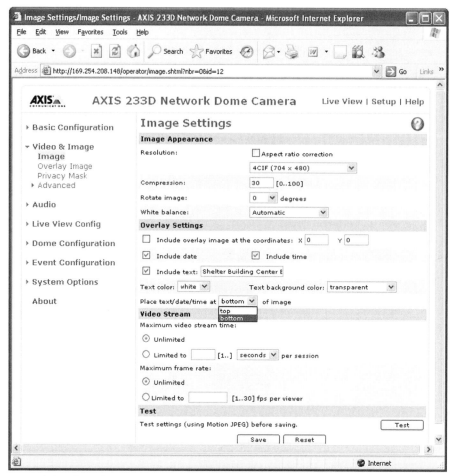

FIGURE 3-23 Web interface of an IP camera.

- Check Include Text (type in the camera name/location in text field).
- Text Color: White.
- Text background color: Transparent.
- Place text/date/time at: Bottom (see Figure 3-23).

iii. Under Video Stream change Maximum Frame Rate to: Unlimited
- Click on SAVE.

PTZ Presets

If the camera installed is a PTZ camera, there may be a specific area of coverage within the requirements designated as the home position or Preset 1, Preset 0, or "Home." This becomes the default if the PTZ was left in some obscure position; the camera can be set to return to the home position after a configured amount of time. If this feature isn't available within the digital video encoder or IP camera interface, check the VMS software.

Resetting to Factory Default

An effective troubleshooting step is to reset the device to its factory default settings. These settings are what passed quality assurance at the manufacturer prior to customizing the configuration for the treated solution. Inside the firmware (the embedded software that runs the hardware) is an isolated section of flash memory that holds the read-only factory default settings. When the device is reset, those settings then overwrite the customized settings, clearing any misconfiguration or corruption that may have occurred during the course of deployment. This may be an immediate troubleshooting step or one of last resort, depending on your familiarity with how the device functions and reacts in certain situations.

Digital Video Cables and Connectors

There are two categories of digital video connectors: those that transmit video and those that display video, and some connectors are used for both. The type of connector used is based on the presence of any analog devices.

Video displays began with analog television, taking a single radio frequency (RF) signal and delivering it to a receiver, originally the old television antennas. When cable television was introduced in the 1940s to better service communities hidden in mountain valleys that were unable to receive television reception, a single signal was transmitted over a coaxial cable (coax) similar to what's used today.

Broadcast coax uses the F-pin connector, frequently seen on the back of most televisions. The RCA composite connector and the BNC twist and lock connector are common to non-broadcast video transmission. Coax is the most common cable used for transmissions of RF frequencies from the 5-MHz analog video signal to the 5.8-GHz signal from a transmitter and/or receiver to an external antenna (see Chapter 5). They're typically classified according to impedance or RG type; for example, a 75-ohm coax or an RG59 type includes two conductors separated by a dielectric material. The center conductor and the outer conductor, or shield, are configured in such a way that they form concentric cylinders with a common axis, which is why they're called "coaxial."

There are also analog cables used for computers, which take the digital image from the video card and convert it to VGA 15-MHz signals (40.7 MHz for XGA) with the two fields of interlaced lines to an analog CRT monitor (much like a television). This traditional 15-pin video graphics array (VGA) connector segregates the RGB signal between inputs and outputs of red, green, and blue on its 15 pins to generate a better picture than the single coax cable, or yellow RCA composite connector, which takes a single analog video signal and transfers it to an analog television monitor. The S-video cable connector has improved that signal by separating the RGB signal into separate pins, as does the 15-pin VGA cable.

This is one of the reasons video games previously looked better on computer screens than on the television, even though they relied on the same 480 lines of resolution (see the section Resolution in Chapter 2). There are also multiple RCA YCbCr connectors called component video cables that raise the level of quality by clearing up the signal for each color. Figure 3-24 shows examples of different video cable connectors.

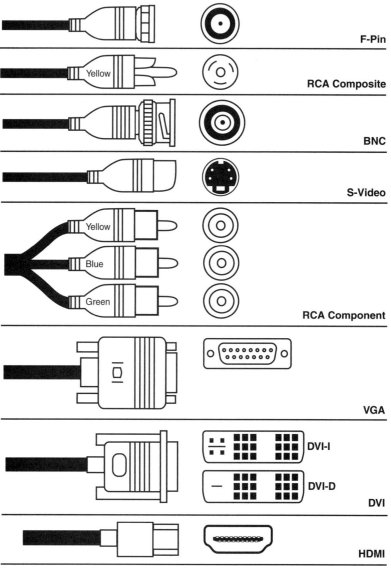

FIGURE 3-24 Video cable connectors.

Now with the convergence of digital video, more and more digital connectors and cables are used to take that analog signal from the cable box, video game console, DVD, and/or Blu-ray device to a digital display. Many LCD displays (HDTV or otherwise) come with multiple options for connections to not only service the many devices in our electronic playground, but to support those devices left over from the analog world of composite video, S-video, and coax cables.

■ ■ ■ ▬▬▬▬▬▬▬▬▬▬▬▬▬▬▬▬▬▬▬▬▬▬▬▬▬▬▬▬▬▬▬▬

Adapters and Baluns

There are many types of adapters (Figure 3-25) that allow you to use existing infrastructure and cabling, including Balun (Figure 3-26), which will convert the signal from a BNC video cable into twisted pair over telephone lines.

FIGURE 3-25 Video cable adapters from left to right: RCA composite female to BNC male, BNC female to RCA composite male, and a female-to-female BNC coupler to extend a BNC cable using another BNC cable.

FIGURE 3-26 A Balun converting the video signal from a BNC cable into twisted pair to run the signal across the building using the existing telephone lines.

■ ■ ■

HDTV and computers are both moving to high-definition multimedia interface (HDMI), as most video graphics cards offer the digital video interface–digital (DVI-D) connector and the HDMI connector. VGA is a purely analog video signal connection to a video card with the 15-pin VGA port to the traditional CRT monitor, which requires an analog signal. DVI (DVI-D) and HDMI, both purely digital transmissions, can receive and deliver more than video data between the video card and the display, providing more support and control over screen resolutions. Digital video interface–integrated (DVI-I) analog and digital is another unique connector that provides both analog (video) and digital (data) transmissions. There's also a DVI-D and DVI-I Dual Link connector for use with multiple monitors.

DVI-D and HDMI are identical in how they transmit data except that the HDMI connection includes encryption data that make it impossible to copy the signal.

Cable Termination

Many times improvisation is required to make a unit work in the field. Ideally, this doesn't have to be the case every time, but it's difficult to anticipate all possible scenarios and obstacles until you're physically at an installation location. This is why making cables onsite is preferable to having them factory-made at specific links. This holds true for all cables, from the Cat5, 5e, or 6 used with IP cameras or to connect the digital video encoder to the network to the coax used to deliver NTSC video to the digital video encoder from the analog camera.

Generally, it's safer to depend on a factory-made cable, with its strict QA process of elimination, but there's a possibility that the cables may not reach (unless a detailed analysis of the conduit pathways used to link the camera and/or digital video encoder was accomplished), especially if the location of the installation changes or becomes more constricted. Cable termination onsite also opens up the possibility of using 90° angled connectors.

Both Ethernet and coax cable termination will typically require the same types of tools. There are also screw- or crimp-type connectors. The screw-types, or "EZ" connectors, are primarily for the coax cables from video transmissions to RF signal transmissions. They literally use the same type of cable only with different sizes and internal construction. RG58 and RG59 can both be used for a video coax cable from, and analog camera to, the digital video encoder, or as an antenna cable from an AP or mesh radio to an external antenna (more on antenna cables in Chapter 5). The EZ connectors are designed to screw onto the end of the cables or act as a locking mechanism. Depending on the type of cable, there's still a cable preparation step required before termination.

■ ■ ■ ▬▬

Ethernet Cable Termination Tips

Tip #1

I've found that my RG9 cable stripper can easily cut the installation of my Cat5e and Cat6 Ethernet cable without damaging the internal cable insulation and copper. When we're out in the field, we all tend to welcome shortcuts so any wire stripper will do, but be careful not to cut into the copper (see Figure 3-27).

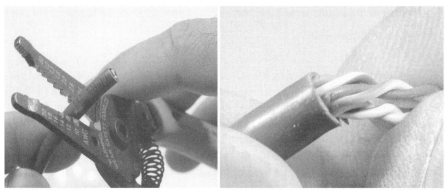

FIGURE 3-27 We all take shortcuts, but avoid damaging the twisted pair when stripping.

Tip #2

Okay, so I use RJ45 Boots. If you've terminated hundreds of cables and are in the field long enough, you'll find that running cables or just pulling them out of a box or vehicle can easily rip the irre-placeable locking pin off the RJ45 connector making the cable useless unless you re-terminate the end. By using these "booties," the locking pin of the RJ45 connector is protected from damage (see Figure 3-28). Once damaged, the connector doesn't lock into place, and the slightest vibration may slide it out of the port, causing communication problems that could've easily been avoided.

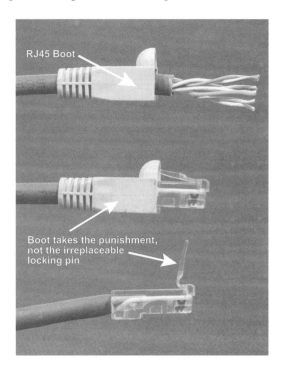

FIGURE 3-28 Boots.

■ ■ ■

Video Coax Termination Tips

Tip #1

The coax cable includes a rubber outer insulation, a braided metal shield, dielectric insulation, and then a single copper wire in the center. That single copper center must be set firmly into the connector tip, because without that connection, or with a poor setting, the RF waves can scatter and cause other interference issues that affect the quality of the signal and thus the video (or wireless link; see Chapter 5).

The nature of the BNC coax connector makes it easy to damage the cable by exceeding the recommended bend radius or forcing the internal male copper end to fall out of the internal female end inside the connector. If you're using the cable in tight places, go with a 90° right angle connector (Figure 3-29), which is specially designed to transmit the signal at an angle.

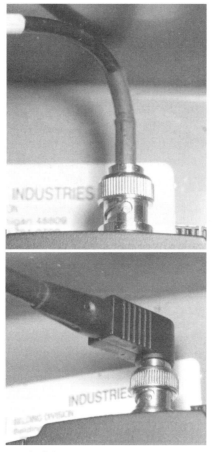

FIGURE 3-29 Go 90 degrees when space is tight.

Tip #2

Coaxial cable is also measured by signal loss per 100 feet. Typically, the thicker the cable, the farther the signal can be transmitted at a minimal loss. Running coax cable from a digital video encoder to a camera 20 feet away makes it possible to use miniature coax cables for ease of installation. But the longer the run, the greater the signal loss, so research the distance and requirements and choose a coax cable that best supports the project. Figure 3-30 shows examples of cable size with associated signal loss.

Belden 9221 Miniature Coax (1.6dB loss)

Belden RG6 Miniature (0.9dB loss)

Belden RG59 (0.9dB loss)

FIGURE 3-30 Coax and signal loss.

DVS Troubleshooting

The more complex the system, the more complex the troubleshooting path and therein lies the "three-dimensional view." Any complex system that includes mechanical, electrical, and software elements requires a three-dimensional thought process for troubleshooting, because what appears to be a mechanical problem may in fact be a software problem. For example, an analog camera connects via coax to a digital video encoder with 18AWG wiring for PTZ control and power. The VMS application is viewed by the end user, who is oblivious to the complexity of the system behind the interface on the monitor in front of him. When a "camera" in the VMS loses connectivity with the VMS, it isn't necessarily the camera that has malfunctioned but the encoder that has the relationship with the VMS system. The problem could be due to camera malfunction or loss of power; encoder malfunction or power loss; Ethernet cables, ports, switches; circuit breaker or power surge; wireless link or radio power; VMS corrupted driver; or changed port allocation. The point is that the problem may not be the camera at all.

The following section presents flowcharts based on field experience and trouble-shooting a number of different installments and configurations.

Troubleshooting an Analog Camera to Digital Video Encoder

When troubleshooting an analog camera connected to a digital video encoder, the three-dimensional troubleshooting view takes effect. Depending on the installation and imple-mentation, the problem could be a mechanical, electrical, or software problem.

How do you know when troubleshooting should be focused on the analog camera and the digital video encoder? There are many variables, but the first step, as determined by the VMS operator, is that there's "No Video." No video could mean many things; first and foremost the encoder doesn't show up in the client software, which is a severed link between the encoder and VMS (software). Here, the first step is to find out if the units are powered. Depending on where the camera is installed, there could be power issues beyond the installation procedures and control, which moves the responsibility else-where. If the building or light pole loses power or is destroyed, there's not much that can be done other than move the unit elsewhere or wait for the power to be re-established by the appropriate maintenance personnel.

If the camera doesn't return online and the digital video encoder reappears in the VMS, then try to log in to the encoder using the Web and/or Telnet interfaces. Some encoders also provide an FTP interface to access the encoder. When the digital video encoder and the VMS software are connected and appear to be communicating correctly, but there's still no video, then the problem is between the analog camera and the digital video encoder.

If the camera and encoder have power, then the next step is a reboot (recycle the power). If the camera returns online and reappears in the VMS, then the next course of action would be to attempt to uncover why the camera went down to avoid the issue in the future. Typically, when a reboot corrects a communication problem between the digital video encoder and the VMS software, there's a software issue that a severed link cleared up to re-establish proper connectivity. If any of these is success-ful and video is present, then it's either an encoder/VMS issue or an issue on the network.

To clarify a clean network connection from the client station to the encoder (bypassing the VMS software), use the Windows PING tool to test the connectivity to the encoder. Bring up the PING tool by navigating as follows: Click START > RUN; then type CMD in the field and press OK. At the command prompt type PING, hit the space bar, and then type the IP address of the encoder. Hit the space bar once again and then type -t (a dash and then the letter "t"). This gives you a continuous PING until you can-cel it by either exiting the window or hitting CTRL C on the keyboard.

If the replies from the unit are in the single or double digits (e.g., Reply from 192.168.1.100: bytes = 32 time = 3 ms TTL = 64), then there's a clean link between the encoder and the VMS. If there are triple or quadruple digits (e.g., Reply from 192.168.1.100: bytes = 32 time = 3867 ms TTL = 64) or a dramatic fluctuation such as

Reply from 192.168.1.100: bytes = 32 time = 3867 ms TTL = 64
Reply from 192.168.1.100: bytes = 32 time = 231 ms TTL = 64
Request timed out.
Request timed out.
Request timed out.
Reply from 192.168.1.100: bytes = 32 time = 4290 ms TTL = 64
Reply from 192.168.1.100: bytes = 32 time = 1235 ms TTL = 64
Request timed out.
Reply from 192.168.1.100: bytes = 32 time = 967 ms TTL = 64
Reply from 192.168.1.100: bytes = 32 time = 825 ms TTL = 64

there may be some network traffic preventing the encoder from staying linked to the VMS. At this point, it may be a good idea to obtain the assistance of the network administrator (see Chapter 4).

A power surge may have damaged the camera and/or encoder, which can be determined if the breaker was tripped or if the power suppressor was damaged (e.g., the unit was hit by lightning).

Software may be the problem, but the most common issue is a security update by Microsoft that replaced a required file or application component, making it unusable by the VMS software, even within their own upgrade, update, or patch. The upgrading of the VMS software may require an update of the device firmware as well. Sometimes it's a corrupted data link that inadvertently dominates the ability of the encoder or the VMS software to acknowledge a handshake, relentlessly sending data packets until the unit crashes or is rebooted to break the cycle. Again, the complexity of a DVS system, using different third-party components, opens the door for incompatibilities and subtle aberrations that never showed up in the QA laboratory.

There are also infallible cable connections to consider – at least, that's the theory. Double check the cables, connectors, and port configurations on all routers and switches. How can cables go bad by just sitting there transferring data? Here are some examples from personal experience:

1. Weather
2. Water
3. Cable fell onto a heater and melted
4. Power surge on switch port
5. Maintenance didn't lock the cable properly
6. A third party cut a series of unrelated cables, but your cable was included
7. Third-party installation of an unrelated implementation unknowingly damaged the cables
8. Poor termination

There are many possibilities, so don't underestimate anything; otherwise it can be a long, frustrating troubleshooting effort. Figure 3-31 illustrates this entire troubleshooting process.

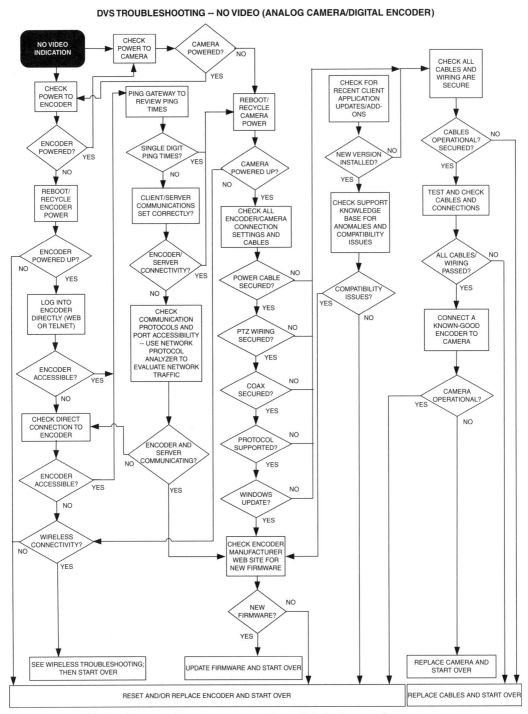

FIGURE 3-31 Troubleshooting an analog camera connected to a digital video encoder.

Troubleshooting an IP Camera

Troubleshooting an IP camera malfunction isn't much different from troubleshooting for an encoder (see Figure 3-32). An IP camera is simply a camera with a built-in digital video encoder. The primary exception in this instance is cabling. An IP camera requires an Ethernet connection to the DVS network, although it still may require a separate power source, if not powered over Ethernet (PoE).

The single 8-pin Ethernet cable not only transmits video data through the network to the VMS software, but also distributes power and PTZ controls, so it becomes the mission critical connection for the operation of the camera and also makes for easier troubleshooting.

Troubleshooting PTZ

Troubleshooting PTZ is unique to each IP camera and/or analog camera and digital video encoder combination (see Figure 3-33). An IP camera has the communication protocols built into its driver so, theoretically, the VMS software would recognize the type of camera and all of its functions. If not, then there are physical mechanical dip switches inside the camera that may need to be repositioned to activate the correct protocols for communications. If that doesn't work, then there may be a compatibility issue, so check both the manufacturer and VMS software Web sites. Since IP cameras are relatively new, the manufacturer's support group may not know how to configure the camera to every software package available having only used a few in their laboratory environment to test. More than once, I've found that I can share information with the technical support staff based on my experiences in the field.

When connecting an analog PTZ camera to a digital video encoder, there's usually a set of dip switches on the camera that provide some control over the type of communication protocol, Baud rate, and serial connection used to communicate with the digital video encoder. These dip switches are an essential part of correctly configuring the camera to communicate with not only the digital video encoder but also the VMS software (see Figure 3-34).

The first step is to make sure that the camera is at least talking to the digital video encoder and with the right Baud rate speed. All of the other elements may be in the correct position and set to the right settings, but without the camera speaking to the digital video encoder at the same speed it's translated as gibberish.

If the analog camera is brand new out of the box, the documentation will contain the factory default setting for the serial Baud rate. If it's a camera that was once used in the field, you may have a problem unless the booting process of the camera is visible within the digital video encoder (usually the boot process displays the basic camera settings, including what serial Baud rate it's using). The Pelco D and Pelco P serial communication protocols are general PTZ communication protocols, even in analog cameras other than Pelco. As an industry leader, Pelco has a large implementation presence and thus, most digital video encoders can communicate using Pelco D and Pelco P. Although

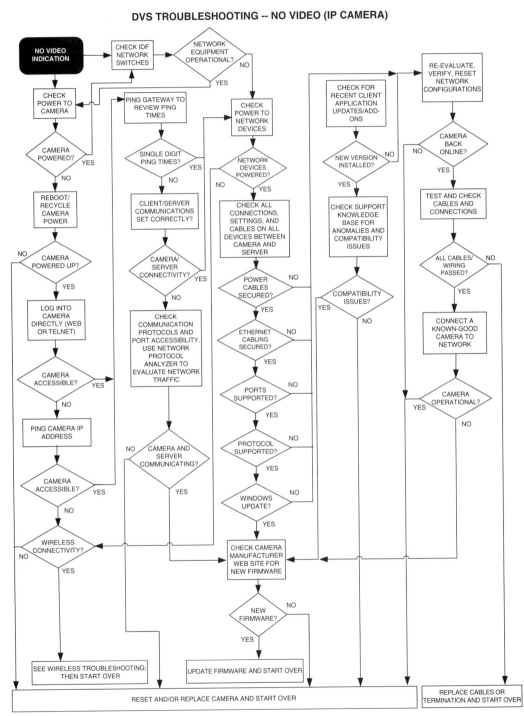

FIGURE 3-32 Troubleshooting for an IP camera.

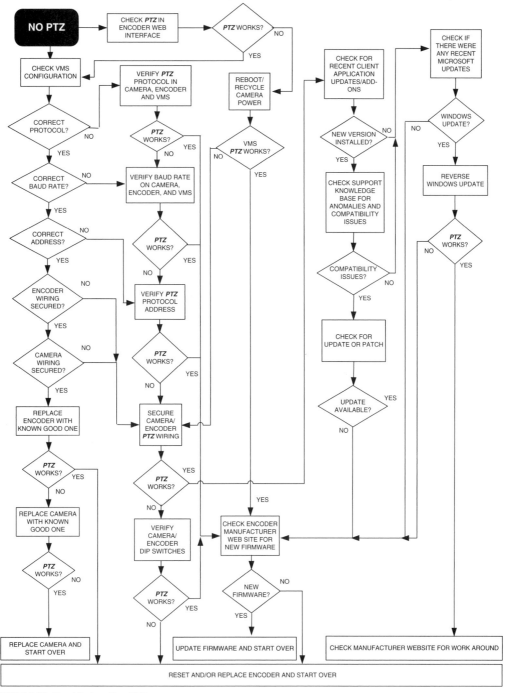

FIGURE 3-33 Troubleshooting PTZ.

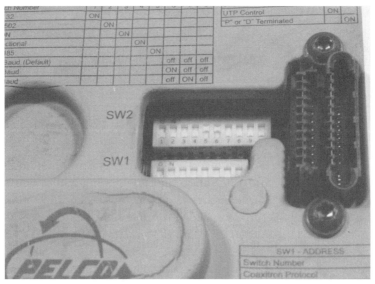

FIGURE 3-34 Dip switches.

configuring a non-Pelco analog camera to communicate with any digital video encoder for PTZ control isn't as straightforward as with a Pelco camera, it can be done with much trial and error. Beware the marketing hype! Stating that the analog camera can be configured for PTZ communications using Pelco D or P could mean between said camera and a single digital video encoder brand name. It pays to actually see it in action, maybe at a live implementation, or test the link in a laboratory before committing to it.

Troubleshooting Poor Video

There are a number of potential culprits for poor video, including wireless connectivity (which is covered in Chapter 5). Assuming a wired connection, troubleshooting an IP camera and an analog camera using a digital video encoder is different only because there's a mechanical connection between the actual physical camera and the physical encoder (BNC coax cable). These tend to be from two different manufacturers adding a layer of complexity. An IP camera has that connection on a printed circuit board, or from one board to another, and it's protected and proprietary.

Poor video falls into this same three-dimensional troubleshooting arena, as there can be mechanical, electrical, and software issues. The first step in troubleshooting should be a reboot or recycle of power to see if that clears up the video. If that option isn't immediately available, then check the resolution, auto-iris, video quality, frame rate, hue and saturation, gamma settings, and method of transmission. Also check the codec used (e.g., MPEG-4, MJPEG, H.264, etc.), as it is possible that someone else changed the settings without knowing the codec isn't supported or wasn't properly configured for its use.

If the reboot is possible, and the video quality returns thereafter, then there's a smaller list to check usually involving the connectivity between the digital video encoder and the camera or the digital video encoder and the VMS.

Upon recycling power, if the quality video returns, the next steps include the basics. Check the BNC cable between the camera and the digital video encoder and make sure it's securely locked in place and there's no damage to the ends or the actual cable. Next, check the connectivity between the encoder and the VMS software, not the video connections but the TCP command connections. Obviously, something happened that caused the encoder to ignore the control commands from the VMS software. Check the ping times for the encoder and determine if there's a networking issue.

If this issue happens frequently, it may be a good idea to check the manufacturer's Web site for a digital video encoder firmware upgrade, which may include a fix. I would like to say that all encoders work seamlessly with all VMS software, but unfortunately that's far from reality. Even the most popular brands of digital video encoders have issues with the most popular brands of VMS software; the technology is always changing and new versions, which may correct one or two issues or add brand new features to stay ahead of the curve, may unintentionally add compatibility problems with last month's digital video encoder model.

One of the better but time-consuming troubleshooting steps is to reset the digital video encoder and/or camera to factory default and rebuild the device from there. Many times the device has difficulty communicating with new VMS software from its current configuration, but when using the factory settings, with less data overhead, the commissioning process is streamlined.

If all else fails, replace the IP camera and/or digital video encoder with a known good one to see if it duplicates the symptoms. If using an analog camera connected to the digital video encoder, then replace the camera thereafter if the symptoms reappear when replacing the digital video encoder. See Figure 3-35 for a troubleshooting flowchart.

Firmware

The software embedded in the memory chip's inside hardware, which controls the hardware, is called firmware. There are two basic frameworks for firmware. One is the type firmly embedded within the hardware and stays there until any upgrade is successful and once successful, the unit reboots and inherits the new firmware. If the firmware upgrade fails, there's no damage to the unit and you can try again.

The other type of firmware doesn't have this backup architecture. If the firmware upgrade is interrupted, such as by power loss, the managing laptop crashes and reboots causing the firmware inside the unit to be corrupted. Hardware without the proper operational firmware is useless.

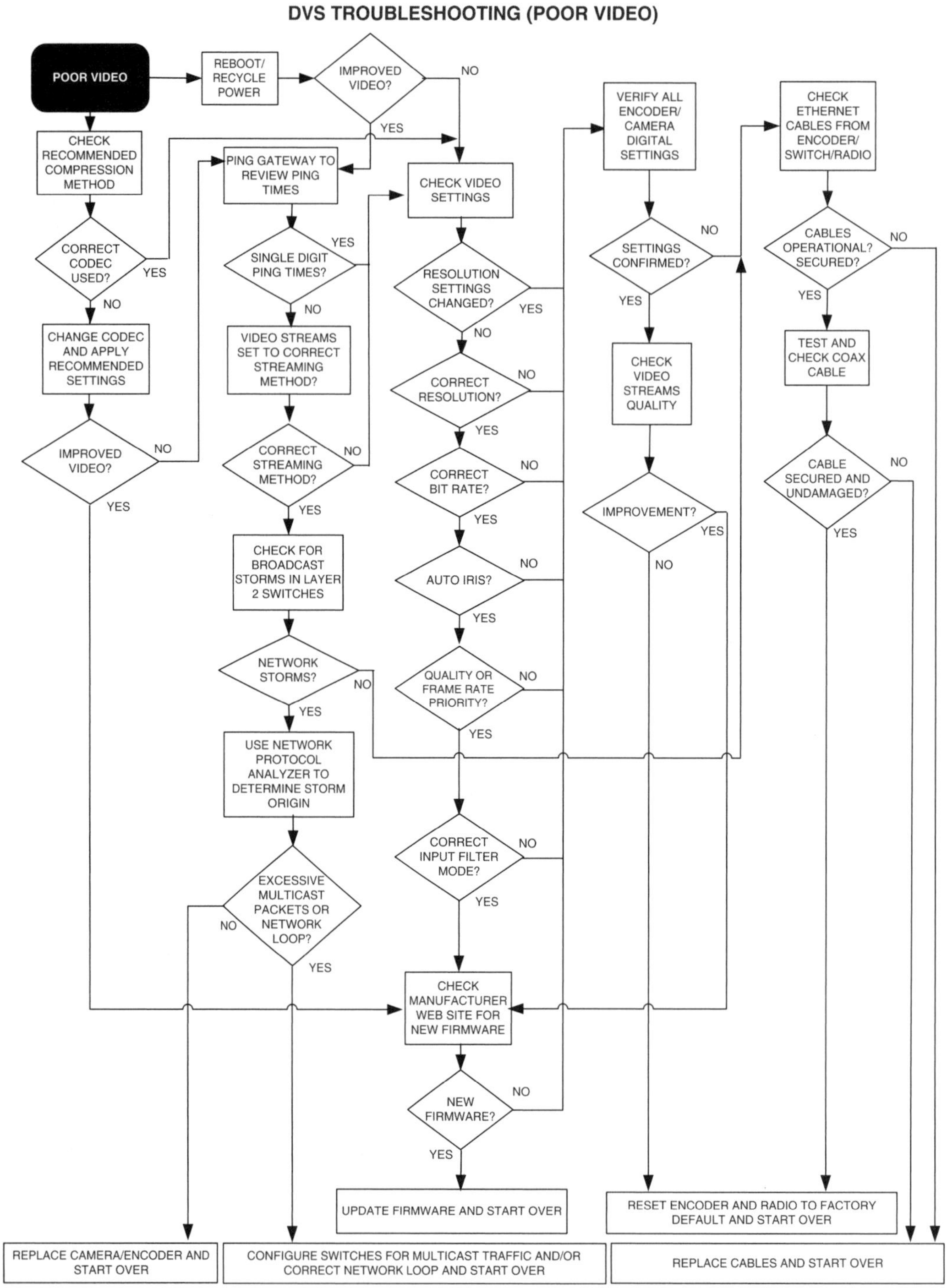

DVS TROUBLESHOOTING (POOR VIDEO)

FIGURE 3-35 Troubleshooting poor video.

Whenever upgrading firmware it pays to make sure there's a solid connection between the managing laptop and the unit, plenty of power for both, and enough time to complete the process, which depending on the device can take as long as 5 minutes.

Troubleshooting Camera Power

The Figure 3-36 camera power troubleshooting flowchart is for an analog camera powered separately (not PoE) and connected to a digital video encoder. This is a unique

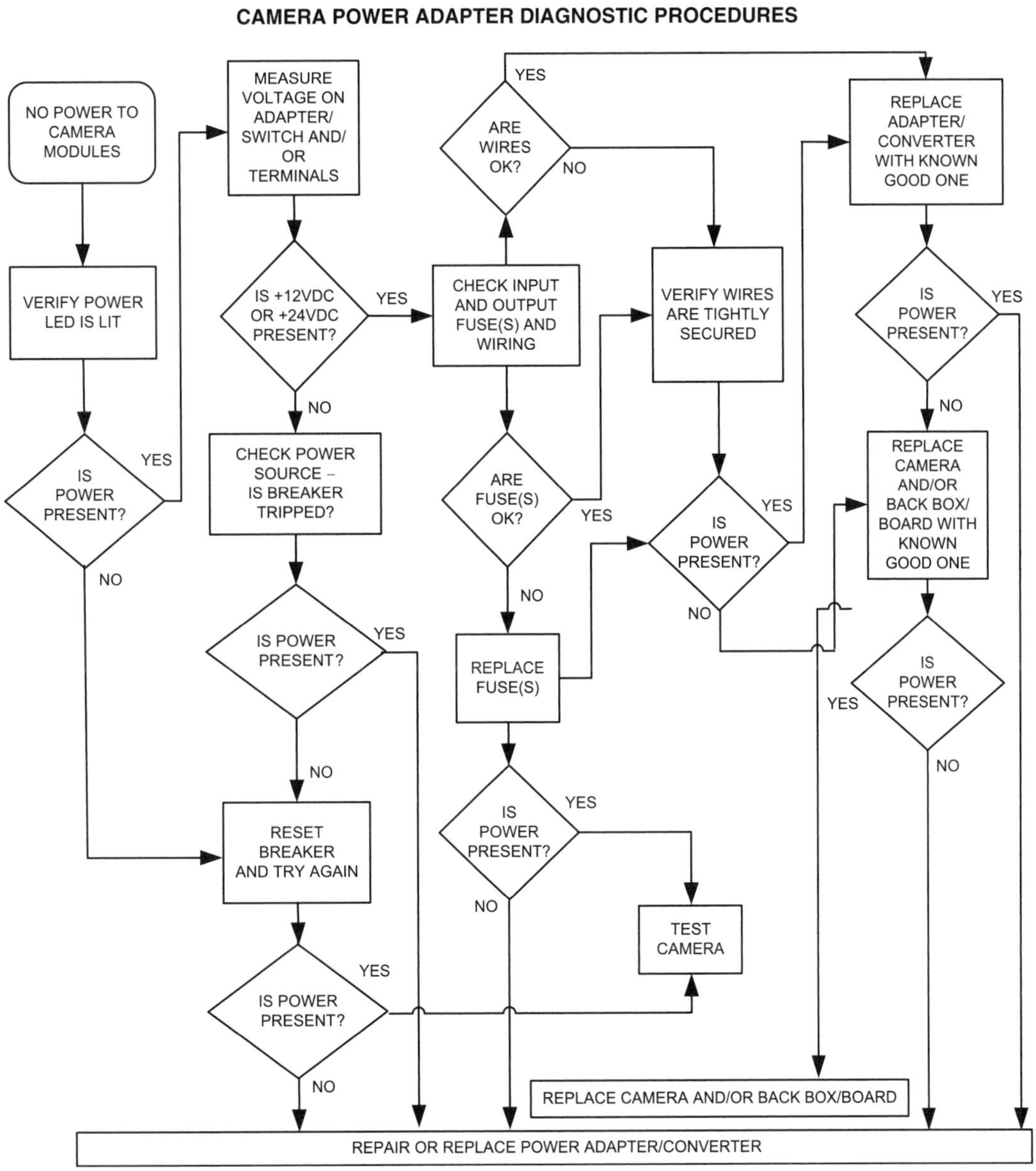

FIGURE 3-36 Troubleshooting camera power.

situation because the digital video encoder is typically listed within the VMS software as the camera, even if the image shows nothing but black, black with the words "No Video," an icon, or another clue that the video connection has been lost. For example, Verint digital video encoders present a half red and half black screen when the camera and the digital video encoder lose video connectivity. When this happens, power isn't the only potential culprit, but it tends to be the easiest to solve.

Each camera has some power LED light that illuminates upon activating power. This LED may not be visible inside an outdoor enclosure, so at least the dome would have to be removed to check. If the LED light is on, then the next step would be to test power. Although I always recommend a power recycle or reboot as the first troubleshooting step, it may prevent you from noticing a degraded or failing power adapter. Test the power leads with a multimeter to determine if there's adequate power. Once power is verified, then move toward a power recycle. If there's poor and fluctuating power coming from the adapter, then it needs to be replaced, preferably with a new one.

There's usually a fuse on the back-box PCB board. This PCB board manages the power, video, and (if applicable) the PTZ controller function. Check to make sure the fuse is good and if not, replace it. If this is the second time the fuse has been replaced, then replace the entire PCB as it may have developed latency failure problems.

Typically, if there's a power problem, it's with the power adapter, the PCB board, or the camera.

Chapter Lessons

A few things to keep in mind when researching a hardware solution:

- With a completely new implementation, work backward from the chosen VMS software solution. Make sure the chosen hardware is fully compatible with the software.
- If there are existing cameras to integrate, choose digital video encoders that fulfill system requirements.
- Research the existing network topology thoroughly with details on all networking hardware and its compatibility with additional IP hardware.
- What environmental factors determine the right hardware for this particular implementation? Weather, lighting, and temperature all contribute to choosing the appropriate hardware.
- Non-automated business processes always have room for improvement, so the smarter the hardware, the more efficient the security personnel will become.
- What are the bandwidth limitations? It would be disastrous to choose ten 5-megapixel cameras with less than 10 Mbps of network bandwidth to deliver all that video.
- Find out where the products have been implemented and research their performance.

4

Understanding Networks and Networked Video

Introduction

The system architecture that uses a server to provide resources to a group of individual computers (called clients, hosts, or nodes) is called a network or networking. The server is usually a high-performance computer that services many groups. Known as an Internet protocol (IP) network, it can change the way people use their computers and software by making it painless to work together as a team or group. It also provides the seamless interconnectivity that's the foundation of e-mail and the Internet, and has given way to digital video security (DVS).

The Power of the Network

Robert M. Metcalfe, inventor of Ethernet, the most common technology for fast local area network (LAN) connectivity, believes that the network creates power or energy when interaction happens between the connected people and/or systems. In his attempt to explain this phenomenon, he describes it as the "Power Law." He claims that the power of any network goes as "2 to the nth" where n is the number of nodes (clients) of a network. Metcalfe's Law (as it's been renamed) explains that the power of the network increases exponentially by the number of computers connected to it by both providing and using resources.

■ ■ ■ ──

The power of a network creates monetary value through cost efficiency. For example, Table 4-1 shows that the "cost" of a long distance call from New York to London in 1927 was 200 work hours compared to only a couple of minutes today. Network power also provides time savings and access to information otherwise unavailable to customers, associates, and partners, thus making it possible to do better business. The computer network, whether it's a LAN or the Internet, has taken this concept to a higher level by interconnecting people, places, and things asynchronously; there doesn't need to be anyone on the other line to complete the connection or to add value to the network.

The more people joined the network, the more cost-efficient the network became.

── ■ ■ ■

Table 4-1   The First Electronic "Network" and How Its Growth
Benefited All Participants

International Telephone Prices Year	Standard Rate (3-minute Call from NYC to London)	
	Current Dollars	Hours of Work
1927	75.00	200
1928	45.00	120
1930	30.00	80
1936	21.00	56
1944	21.00	40
1945	12.00	20
1969	12.00	6
1970	9.60	5
1974	5.40	1.6
1980	4.80	0.9
1986	4.83	0.7
1991	3.32	0.3
1995	2.40	0.2
1999	0.30	0.02

Networks serve to bring together people with the best tools to accomplish the best work. Computer networks have given us:

- E-mail
- File and print sharing
- Shared hardware resources
- Internet access
- Firewalls
- Web servers
- FTP servers
- Remote access to applications
- Scalability
- Integration of different departments and systems
- Video conferencing
- Chat
- Digital white-boarding for remote collaboration
- Global backup and recovery

Not only do networks provide integration of different departments and systems, but they also provide the translation of multiple devices and software environments, creating better connectivity beyond the Cat5 cable. The client/server programs become the translator over the Ethernet network. For example, any workstation, as long as it's connected into a TCP/IP network, could be using any number of operating systems (OS), but as long as it's using a standard client (Web browser) it can connect to any server (Web server) on any machine in the world – no matter the language, OS, or country.

Business opportunities can be lost when resources and knowledge remain hidden or unshared. Relative to this book, you may be unaware that real-time video surveillance can be accessed anywhere inside or outside the office and can decentralize security for a more mobile system. It's important to uncover the collaborative tools that bring cohesiveness to your company. An integrated DVS system can be one of them.

Getting Wired

The Open System Interconnection reference model (OSI) by the International Standard Organization (ISO) is a primary architectural model for communications between computer systems. Many of the components of the OSI model exist in all communications systems. It defines the typical networking system architecture in seven layers: application, presentation, session, transport, network, data link, and physical layer. Each has its own unique functions and protocols. A computer networking protocol is the set of standards and rules that hardware and software must follow to communicate with all other computers. Transport control protocol/Internet protocol (TCP/IP) has become a standard for computer communications. Any device running TCP/IP can communicate with another shared device, anywhere in the world – running any type of OS, software, or network topology – as long as it's also running TCP/IP. The four general layers that make up this "translator" TCP/IP (or Department of Defense; DOD model) include:

- Process and application layer
- Host-to-host transport layer
- Internet layer
- Network access layer

The application layer holds the actual interface to the software and utilities that allow networking functions such as file and print sharing and authentication (e.g., HTTP, FTP). The application layer passes the newly created data to the transport layer. This layer consists of the transfer control protocol (TCP) and the user datagram protocol (UDP), and is all about traveling from one place to another (Figure 4-1).

The Internet layer includes the IP, address resolution protocol (ARP), and Internet control message protocol (ICMP). IP addressing standardizes how computers communicate with each other. This core scheme is responsible for the ubiquitous communications between devices that have made convergence much easier.

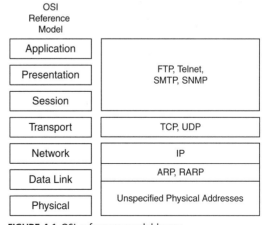

FIGURE 4-1 OSI reference model layers.

An IP address contains four octets of decimal numbers derived from binary logic. IP is simply the protocol used to fragment packets and provide logical communication. For example, domain name server (DNS) on the application layer will resolve 192.168.0.1 to www.websitename.com. Address Resolution Protocol (ARP) will then resolve the IP address to a media access control (MAC) address on the data link layer (or network access on the DOD) and Reverse Address Resolution Protocol (RARP) will do the reverse.

The lowest general layer is the interface (physical) layer, which consists of the data link and hardware used for connectivity. In this layer Ethernet joins the equation and the data link. Ethernet is possible with the inclusion of specific hardware, which is the physical hardware component of the layer. The interface layer also provides the error-checking and correcting functions: checking for errors inside incoming data or packets, adding error-checking information to the outgoing data, and resending the data if there's no returned acknowledgment of delivery.

The network interface card, sometimes referred to as a NIC (pronounced "NICK") card or network adapter, is another physical component of this layer. Along with its unique physical address or MAC address, it provides a unique signature on every network adapter, thus giving each computer a unique identification. The remaining majority of this layer is invisible, greatly reducing its complexity. The physical MAC address of a computer's network adapter can be found by using the IPCONFIG/ALL command at the command prompt.

START > RUN > Type CMD and press ENTER

Once the command prompt appears, type IPCONFIG/ALL (see Figure 4-2) and press ENTER. The MAC address is called the physical address and is a series of double-digit octets separated by colons.

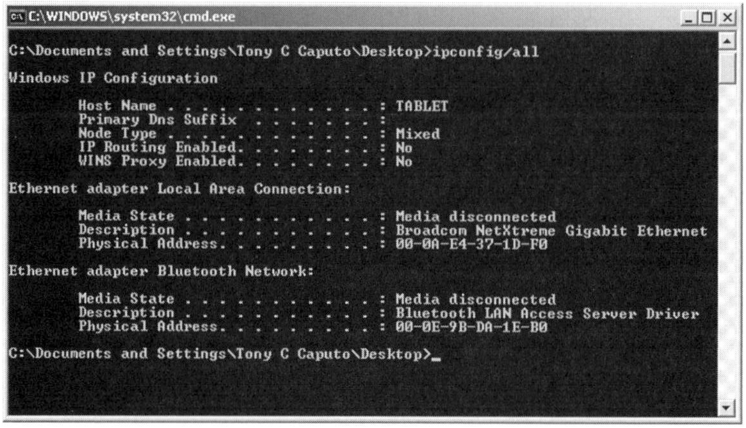

FIGURE 4-2 To find your MAC address, at the command prompt type IPCONFIG/ALL and hit ENTER.

Why Ethernet

There are a few reasons why Ethernet has become the networking juggernaut. First and foremost Ethernet equipment can be very cost-effective. The second reason is that Ethernet is infinitely extensible, managing the ever-demanding traffic of data, voice, video, and storage. Ethernet innovates to more easily evolve with its ever-growing bandwidth demands, beginning at 10-100 Mbps and onward to 1000 Mbps, 10 Gbps, 40 Gbps, and even 100 Gbps Ethernet.

The third reason is familiarity. Even when companies or associations introduce new, exciting networking technologies, Ethernet is already a comprehensive solution that's easy to understand and actually works. The age-old familiarity with unshielded twisted pairs (UTPs) has proven to make Ethernet an easy and ubiquitous solution. Just give a few Ethernet cables, a $30 switch, and a few client stations to a 10-year-old who wants to play networking gaming and you'll have a functioning network in less time than it takes to boot up Windows.

Only a few Ethernet topologies are used. The original bus topology used a straight daisy-chain of computers in a linear fashion. This limited the amount of distance and the maximum amount of nodes (see Table 4-2). Even the type of coaxial cable used can limit the distance in a linear bus architecture and because there are no hubs, switches, or routers, terminators were required at each end of the daisy-chain.

Table 4-2   General Network Designs

Topology	Ethernet	Description	Cable Type	Maximum Cable Length (m)	Maximum Nodes	Data Rate (Mbps)
Bus (linear)	Yes	Straight daisy-chain connectivity with no hubs or routers, but require terminators at each end	RG8 coaxial (BNC) cable with terminators	500	30	10
Bus (linear)	Yes	Straight daisy-chain connectivity with no hubs or routers, but require terminators at each end	RG58 coaxial (BNC) cable with terminators	185	30	10
Token ring	No	A constant messaging around a ring to avoid data collision	Cat2 or 4	45	72	4 or 16
Star	Yes	Multiple workstations connected to a hub or switch	Cat5, 5e, or 6	100	1024	10/100/1000
Tree	Yes	Multiple interconnected star networks	Cat5, 5e, or 6	100	1024	10/100/1000

The once popular token ring topology wasn't Ethernet, but created a constant messaging around the ring to avoid collision. Token ring used coaxial, Cat2, or Cat4 wire with up to 72 nodes (more than double that of a bus topology). However, the ring included a maximum distance of only 45 m and while the bus network offered up to 10-Mbps bandwidth, the token ring varied between 4 and 16 Mbps, depending on distance, cabling, and architecture.

The star and tree topologies are true Ethernet architectures, providing speeds of up to 1 Gbps (more with the upcoming 10-Gbps Ethernet) and up to 1024 nodes. Although Cat5, 5e, or 6 cable is the standard cabling with the RJ45 connector for Ethernet and it's limited to a distance of 100 m, that distance is between each network device. For example, a computer can be 1000 m away from a network router, as long as there are two intermediate network switches at 100-m intervals.

If you're using Windows as your workstation OS, you can create a simple peer-to-peer (serverless) network by installing network adapters into a free PCI slot on each of two computers (if not already built into the motherboard) and installing the appropriate software driver. Depending on the age of the NIC card, the drivers may already be part of the Windows OS or may come with the network adapter on a CD-ROM. You can connect the two computers by plugging a Cat5 crossover cable into the RJ45 Ethernet port on each network adapter. The two computers can now communicate using the TCP/IP built into Windows by way of the network adapter. The same holds true when the same Cat5 cable from a computer is plugged into an IP camera or encoder, both of which include an embedded OS that also uses TCP/IP.

A crossover cable is used to interconnect two computers by "crossing over" their respective pin connectors. Once the cables are plugged in, the two computers start communicating with each other using TCP/IP and a successful simple peer-to-peer network is created (see Figure 4-3).

Peer-to-Peer and Client/Server

Peer-to-peer and client/server are two basic network types. Peer-to-peer interconnects independent workstations to share files and printers (see Figure 4-4). Creating a peer-to-peer network is an easy installation, but it drains resources from each individual workstation, thus slowing down performance. Peer-to-peer is also less secure than a server-based network.

The client/server environment provides a multitude of advantages I've touched upon throughout this book, including:

- Better security
- Centralized administration
- Reduction of processing burdens on individual workstations

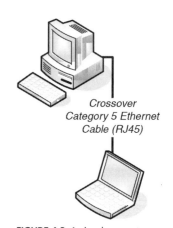

Crossover
Category 5 Ethernet
Cable (RJ45)

FIGURE 4-3 A simple peer-to-peer network: connect the Ethernet ports together with a crossover cable.

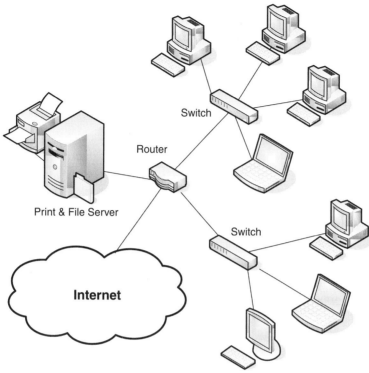

FIGURE 4-4 Basic network sharing.

- Increased collaboration
- More extendability
- Better maintainability

The first step in network design involves creating a plan for configuring and arranging your network, also known as a topology (see Figure 4-5). A topology is a pictorial description of how the network is set up, driven mostly by the office environment and locations. There are three basic types of topologies: bus, token ring, and star. The bus topology is a daisy-chain linkage with termination and ground required on both ends. The token ring network incorporates nodes within a circular channel that's always "on" and looking for the right data. The star topology is the most popular, because it gives you the option of linking as many as 1024 nodes to one network, whereas the bus network holds a maximum of only 30 nodes, and the token ring holds up to 72 nodes.

Ethernet Equipment

There are a number of special devices that have matured over the years and provide select functionality to better the load and operation of your IP network. The following is a summary of each device that will be discussed in detail throughout this book.

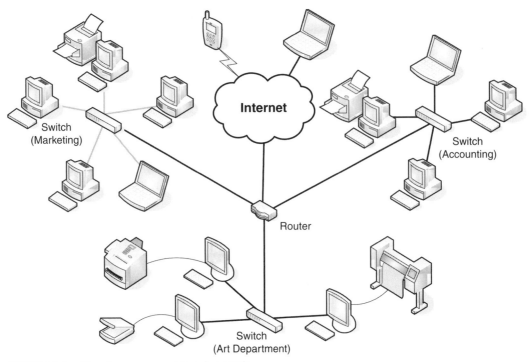

FIGURE 4-5 Small/medium company network example.

Router

A router is a hardware routing device. A router works at Layer 3 of the OSI model – the network layer – and thus can communicate between different networks. Layer 3 is where the IP protocol examines the source and destination of packets between IP addresses. In the event that the destination isn't listed, the router will either send the packet to a default router or drop the packets as lost. Routers are usually used to connect a local area network to a wide-area network (a LAN to a WAN) because routers prevent broadcast traffic from bleeding between networks by isolating their own broadcast domain. If the network is bombarded by IP broadcasts, the router allows you to segregate traffic using subnetting. Only the router allows broadcast traffic to flow between subnets. Because routers prohibit broadcast traffic and give you control over the TCP and UDP ports, they make exceptional firewalls.

Switch

A switch is a hardware device that works at Layer 2 (data link) of the OSI model. The data link layer is where the Ethernet protocol works. A switch "switches" Ethernet frames by keeping a table of which MAC addresses have been seen on each switch port. The switch uses this table to determine where to send all future frames it receives. In Cisco terminology, this table is called the content addressable memory (CAM) table. If a switch receives a frame with a destination MAC address not already in its table, it floods that frame to all switch ports. When it receives a response, it puts that MAC address in the table so that it's recognized and the switch won't have to flood all ports next time it appears. A switch might be considered a "smart switch," which only means it gives you added functionality, including remote manageability, port management, status, service logs, virtual local area network (VLAN) configurations, spanning tree, and MAC address level security controls.

Hub

Hubs are interconnected hardware devices with a minimum of four Ethernet ports that send signals (packets) of data from one port to all other ports, creating a continuous connection between devices. A hub is a multiport repeater, so any data that come into one port of a hub are duplicated and sent out to all other ports with attached devices. There's no intelligence in how a hub functions, and they're not recommended for heavy data transmissions. Avoid hubs within the network topology.

Many of today's Ethernet hubs and switches work at dual speed: the original 10BaseT (10 Mbps) and the newer Fast Ethernet or 100BaseT with speeds of up to 100 Mbps, with newer models providing gigabit connectivity (1000 Mbps). The "10," "100," or "1000" in the name refers to the transmission speed of either 10, 100, or 1000 Mbps, respectively. Base refers to an exclusive Ethernet baseband signal and the "T" designates the media, which carries the signal (twisted-pair). It's important to keep speed in mind when purchasing a hub or switch. The speed needed should be determined by the bandwidth requirements. There's a noticeable difference between data transferred on a LAN at 1000, 100, and 10 Mbps, especially when adding real-time streaming video 24/7. Whether you're designing a network or extending one, an existing, detailed network topology with all model numbers and configurations of switches, hubs, and routers is a must.

If a DVS system will be used by select personnel at various buildings on campus, all of which are linked via fiber-optic cable and gigabit switches, the addition of a 100-Mbps switch will become a bottleneck, and a 1000-Mbps switch should be considered.

Firewall

A firewall is typically used to protect your private LAN from the Internet at Layers 3 (network layer: IP), 4 (transport layer: TCP and UDP), and 7 (application layer). There are different types of firewalls, but the most popular type is a stateful packet inspection (SPI) hardware firewall. An SPI hardware firewall is a dedicated device that understands the different states of the TCP protocol, so if a packet attempts to gain access to the network behind the firewall but wasn't requested, the firewall knows and drops it.

I can't emphasize enough the need for a firewall in any IP network design that opens up the LAN to a WAN (or Internet). As an experiment, I installed and configured three different machines and connected them, via a switch, directly to the cable modem. One machine had a default installation of Red Hat Linux 7.1, the second had a default installation of Windows 2000 Server, and the last was an Apple Workgroup server that had a default install of Apple Mac 8.0. I intentionally set this up to determine how vulnerable and accessible a computer was when exposed to the Internet without a fire-wall. The result was that the Windows 2000 Server crashed from a virus within 90 minutes (the vulnerability was corrected in Service Pack 2). The Red Hat Linux crashed after 7 days and the old Macintosh was never hacked into or infected. Although later, when I believed the old Mac to be invincible and found a safe machine to use as an e-mail server without the high maintenance usually involved with protecting a simple mail transport protocol (SMTP) server, the Mac was hijacked within a few weeks – even behind a firewall. Don't underestimate what a hacker, cracker, or computer farm can do while moving in nanoseconds.

Cabling

Table 4-3 lists the common categories of UTP cabling (see Figure 4-6), their bandwidth limitations, and typical applications. Cat5, 5e, and 6 are used for Ethernet networks, using the RJ45 connectors, but only 5e and 6 can be used for gigabit (1000BaseT) connectivity. Tables 4-4 and 4-5 depict the correct method for terminating Ethernet cables, both as Class A and the most common Class B, respectively. Table 4-6 explains how to make a crossover cable.

Table 4-3   Twisted-Pair Copper Cable Categories

Category	Data Rate (Mbps)	Typical Application
1	<1	Analog voice – plain old telephone service (POTS), doorbell
2	4	Mainly used in older token ring networks
3	16	Voice and data on 10BaseT Ethernet
4	20	Rarely used except for 16-Mbps token ring
5	100	100 Mbps (100BaseT)
5E/6	1000 (4 pair)	Gigabit Ethernet

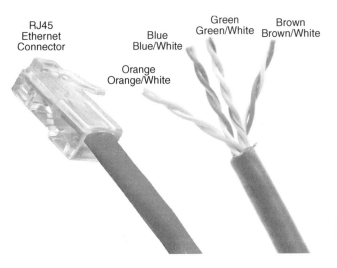

FIGURE 4-6 The standard RJ45, Cat5 (5e, 6) Ethernet cable connector and the reason it's called "twisted pair."

Table 4-4   Standard Ethernet, Straight-Through Class A

RJ45 Pin #	Wire Color (Class A)	10BaseT Signal 100BaseTX Signal Category 5, 5e, 6	1000BaseT Signal Category 5e, 6
1	White/Green	Transmit+	BI_DA+
2	Green	Transmit−	BI_DA−
3	White/Orange	Receive+	BI_DB+
4	Blue	Unused	BI_DC+
5	White/Blue	Unused	BI_DC−
6	Orange	Receive−	BI_DB−
7	White/Brown	Unused	BI_DD+
8	Brown	Unused	BI_DD−

Table 4-5   Standard Ethernet, Straight-Through Class B

RJ45 Pin #	Wire Color (Class B)	10BaseT Signal 100BaseTX Signal Cat5, 5e, 6	1000BaseT Signal Cat5e, 6
1	White/Orange	Transmit+	BI_DA+
2	Orange	Transmit−	BI_DA−
3	White/Green	Receive+	BI_DB+
4	Blue	Unused	BI_DC+
5	White/Blue	Unused	BI_DC−
6	Green	Receive−	BI_DB−
7	White/Brown	Unused	BI_DD+
8	Brown	Unused	BI_DD−

Table 4-6   Ethernet Crossover Cable

RJ45 Pin # (END 1)	Wire Color	RJ45 Pin # (END 2)	Wire Color
1	White/Orange	1	White/Green
2	Orange	2	Green
3	White/Green	3	White/Orange
4	Blue	4	White/Brown
5	White/Blue	5	Brown
6	Green	6	Orange
7	White/Brown	7	Blue
8	Brown	8	White/Blue

Table 4-7   Fiber Modes

Fiber Mode	Distance*	Color Code
Single mode	Up to 100 km	Yellow
Multi-mode	Up to 2 km	Orange

*Depends largely on fiber-optic transceiver.

Fiber Optics

Fiber optics is preferred in high-end digital video surveillance implementations for the bandwidth, usually somewhere between 20 and 40 Gbps. Traditional Cat5 copper can provide up to 100 Mbps and with Cat5e or Cat6 you can reach 1000 Mbps (1 Gbps), but fiber optics (in single-mode) has reached upwards of *140 Gbps*, and TeleKom Malaysia and Verizon Business are working on *213 Gbps* by 2011. See Table 4-7 for different fiber modes.

Fiber can be run in bundles, which makes sense as it's really not the price of the fiber itself that makes it cost-prohibitive, but the installation process. When installing fiber, it must be completely shielded from any power cabling, so best practice recommends running fiber within its own *innerduct*, not only shielding it from all other cabling that may exist within the duct or conduit, but also to better recognize the run later. It costs just as much to install a single strand of fiber as it does a 24-strand bundle.

There are a few different connectors used to terminate fiber (Figure 4-7), which isn't nearly as simple as terminating copper wire. For one thing, fiber can break, it can melt, or it can create latency problems that may not appear for weeks (or until someone happens to bump into it), so this task is best left to professionals.

Ted Gary, Network Architect for Operation Virtual Shield, Chicago's homeland security fiber infrastructure for video surveillance, believes fiber-optic technology is the best thing to happen to communications since the invention of the radio. "Radio allowed concurrent transmission of independent program streams and rapidly expanded the number of applications and users of this base technology," Gary stated. "Fiber optics has done exactly the same thing; with military, commercial and consumers benefiting from this base technology" (personal communication, July 25, 2009).

Gary explained that as radio frequency (RF) progressed from its analog roots to ever more complex and efficient modulation and encoding schemes on

FIGURE 4-7 SC fiber connector on multimode (orange) fiber-optic cable.

the carrier frequencies (in the 1 MHz-300 GHz range), fiber optics was able to use its dramatically higher carrier frequencies (100,000-400,000 GHz) with inefficient encoding schemes. Nevertheless, the experience in RF (and satellite and cable TV) continued to provide strategies for containing the cost of accommodating the exponential growth in data transmission demands over fiber-optic networks. Subcarrier multiplexing, once exclusively used in RF transmission technology to squeeze more capacity out of existing frequencies, is now used in fiber-optic technology to allow lower cost encoding and decoding elements to carry more information within existing single-mode fiber.

Gary concluded by adding that optical packet switching, with roots in the wired networking world, makes the required speed of hundreds of Gbps now viable by standing on the shoulders of these advances in networking.

Broadband over Power Lines

Broadband over power line (BPL) is a relatively new digital communication technology that uses electric power lines (medium and/or low voltage) to transmit and receive data (see Figure 4-8). BPL uses the shortwave frequency spectrum (1.7-34 MHz) to send electronic pulses through existing power lines. This becomes an exceptional solution for old and new buildings where wiring for Ethernet and/or fiber is too cost-prohibitive. In most implementations that require multiple locations with larger buildings, running conduit as per city ordinances and codes is the most expensive portion of the entire project. BPL offers a quick deployment solution by using the existing electrical

FIGURE 4-8 Using broadband over power lines speeds deployment including the mobility of elevators and areas where there is no wireless signals or data raceways.

infrastructure. Of course, if there's no electrical infrastructure, then there's no power and where there's no power, there will never be any cameras.

Multi-dwelling units (MDU), because of their general architecture, offer in-building connectivity through the electric lines because the units receive electricity from a centralized location as long as the overall digital communication is accessible to all electric phases. If the electrical lines are on different phases or circuits, then the digital signals may lose power or won't be available to all occupants linked to the separate circuit.

Thomas Yang, lead architect at IBM Global Services, designed the BPL implementation for the New York Housing Authority and in-elevator surveillance at New York's Trump Tower. Yang explained that the primary challenge in the NYC Housing Authority project wasn't the equipment, but rather choosing the right electrical injection locations for the best BPL signal distribution (again, all about bandwidth). "The site survey and electrical wiring diagrams are critical information for the design and deployment," said Yang. "Noisy or broken electronic devices, like microwaves or light ballast, can introduce serious RF noise into the electrical circuit and harms the BPL transmission" (personal communication, March 3, 2009). Yang also cautioned that old wiring or risers can reduce the range of BPL signal transmission and it quickly deteriorates the digital signal.

One of the best applications for BPL is mobility. New York's Trump Tower used BPL when implementing video surveillance within elevators. Running new traveling data cabling (cabling that needs to move with the elevators) is costly in both implementation and maintenance. By using BPL over the existing traveling power cables, video surveillance streams were able to reach the DVR with significant savings in both time and cost.

As with any digital video security (DVS) project, the client's requirements (areas of coverage, resolution, and frame rate) will drive the solution. BPL and overall network design may only meet location requirements, as injection points may not meet bandwidth requirements, but it's still worth considering as a cost-effective option to Cat5, fiber, or wireless. By using multiple silo injection points and gateway zones, BPL uses the existing electrical infrastructure to deliver the data, so there's no need to run wires (for device power or data) and code-specific conduit. No matter the size of the project, a requirement for extensive runs of code-specific conduit (e.g., parallel coated 1″ PVC for power and data for outdoors) can be a significant part of the project costs.

Corinex, one of the leading manufacturers of BPL equipment, also released a simple plug-and-play device by plugging both modules into a power outlet at each end and pressing a button to synchronize the signal. Once synchronized, the modules include RJ45 Ethernet ports that provide up to 200-Mbps bandwidth across the electrical wiring. The most difficult part of a BPL solution, from my experience, is receiving accurate and current electrical survey and wiring diagrams. As Yang points out, "the suitability of the in-building BPL system to a DVS application is contingent on understanding the electrical infrastructure and that a proper RF site survey is conducted to uncover any electrical interference."

Setting up a Star Network

The most important advantage of the Ethernet star topology is expandability. As long as a computer has an Ethernet port, you can join up to 1024 nodes to a star network by plugging into an RJ45 port on a hub or switch, which may be uplinked from another hub or switch. There's some configuration involved, but with the right OS and server configuration it could be minimal. The requirements for an Ethernet network include compatible OS, network adapters with appropriate drivers, an Ethernet cable for each node, and enough hubs and/or switches to accommodate each connection. If you're planning to link a few dozen computers within the near future, segregate them by department or location. Each group could have their own switch, which in turn can link to another switch, turning your star into a spanning tree topology. This is by far the best course of action as it gives you the flexibility to expand: more nodes, more stars.

IP Addresses

Ethernet networking requires some configuration, implementation, and equipment. A simple method of seeing how TCP/IP works is by linking two computers together to communicate and share resources.

Once you've chosen two machines to network, you'll need to give each machine a unique IP address. IP addressing is a unique 32-bit logical address divided into two main parts, the network number and the host number, and is part of the network layer protocol. An IP address follows a simple format and can be subdivided and used to create addresses for subnetworks. Every host within a TCP/IP network is assigned its own unique IP address. There can't be two hosts within a single network with the same IP address as this will create confusion within the network. Imagine if you were a visitor to a new city and noticed there were two completely different hotels with the exact same address – you'd be confused, too!

To assign an IP address to two machines so they may communicate via TCP/IP, do the following:

1. Install a NIC in each machine (if not already built in).
2. Configure the proper driver on the NIC for the workstation's operating system.
3. Configure TCP/IP by assigning a unique IP address on each node with a common 24-bit subnet. For example, assign node 1 the IP address of 10.0.0.1, with a 24-bit mask 255.255.255.0. Configure the other workstation as 10.0.0.2 with the same 24-bit mask. This will allow them to communicate via TCP/IP.
4. Connect both workstations via a hub, switch, or a crossover cable.
5. Configure your Windows OS to be a client on a network.

You can accomplish this by right-clicking MY COMPUTER > PROPERTIES > NETWORK IDENTIFICATION. Change both work groups to be identical or they won't be able to share resources.

IP Address Format

The 32-bit IP address is grouped in four 8-bit octets, separated by dots and represented in dotted decimal notation. Each bit in the octet has a binary weight (128, 64, 32, 16, 8, 4, 2, 1) with a minimum value of zero and a maximum value for an octet of 255. Figure 4-9 illustrates the basic format of an IP address.

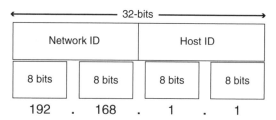

FIGURE 4-9 Anatomy of an IP address.

Also referred to as a dot address, it includes four octets of 8 bits presented as three or fewer decimal digits separated by periods. For example: 192.168.100.100 or, in 32-bit binary language,11000000.10101000.1100100.1100100.

The network identifier and host identifier within the IP address designate its place within that network. Both of the previous examples are the same IP address: 192.168 is the network identifier and 100.100 is the host identifier. If you split the network into subnet groups the network identifier stays the same, but the subnet is identified by the first 100 octet (or 1100100), and the last digits, or second 100, recognize the device. Both the subnet and the device numbers thus become the host identifiers.

The Network Identifier

The Internet is comprised of many individual networks. IP is the method through which one of these networks communicates with another. For this communication to occur your host must have a valid and unique public IP address assigned through an ISP or a numbering authority like Network Solutions (www.netsol.com).

The Host Identifier

In addition to the network identifier, the devices or host machine need identification as sender or receiver. Part of this host identifier signifies the subnetwork using a subnet address, which divides the physical networks into subcategories to handle many devices. The host identifier refers to the remaining numbers available after you subnet the IP address. For instance, if the network is 192.168.0.0 with a 24-bit subnet mask (255.255.255.0), then you can have up to 254 usable network host addresses. When you click on a hypertext link on a Web page or send e-mail, the IP address becomes the address of the sender and receiver.

Domain name service (DNS) is the application service that translates the IP address into a more recognized and memorable name. Whenever using the Internet, there are millions of DNS servers that translate any uniform resource locator (URL) typed into the location field of any Web browser into a specific IP address. Every Web site has a unique IP address. This is easily uncovered by using the Windows command line PING tool (See the Software Troubleshooting Tools section), typing the following, and then pressing ENTER:

C:\>ping www.whateveraddress.com

The reply would look something like this

Pinging www.whateveraddress.com [209.191.XX.XX] with 32 bytes of data:
Reply from 209.191.XX.XX: bytes = 32 time = 42 ms TTL = 52
Reply from 209.191.XX.XX: bytes = 32 time = 42 ms TTL = 52
Reply from 209.191.XX.XX: bytes = 32 time = 42 ms TTL = 52

along with additional statistical information. When resolving to a domain, DNS is what turns the numeric addresses of four separate three-digit-octets into a URL such as www.yahoo.com, which is easier to remember than a series of arbitrary numbers.

IP Address Classes

There are four different address formats, or classes, but only three are significant in a corporate setting. Each class provides for different networks and available hosts according to their size:

- Class A: large networks with many devices
- Class B: medium-sized networks
- Class C: small networks (less than 254 devices)

Anything outside your internal LAN environment needs a formal application to a network solutions authority for a network IP number.

The first few bits of each IP address indicate from which address class it originates (see Table 4-8). The Class A Network begins with a 0. Any binary IP address beginning with zero belongs to a Class A network. The Class B network begins with 10 and the Class C network begins with 110. For example, the IP address 66.218.71.198 belongs to Yahoo.com. Its binary number is 1000010.11011010.1000111.11000110, which puts it into the Class B network category because it starts with 10.

Most DVS applications are, for security reasons, built inside a closed Class C or B network. An additional method is used for remote access, usually using HTTP (port 80), Windows Remote Desktop (port 3386), VNC (port 5001), or a similar application. When using any of these applications to access internal (as in a closed business

Table 4-8 IP Address Bits and Bytes

Class	Initial Byte (First Octet)	First Bit	Network Bits	Host Bits	Multicast Bits	Number of Networks	Maximum Number of Hosts
Class A	0-127	0	7	24	N/A	126 (0 and 127 are reserved)	16,777,214
Class B	128-191	10	14	16	N/A	16,384	65,532
Class C	192-223	110	21	8	N/A	2,097,152	254
Class D	224-247	1110	N/A	N/A	128	See RFC 1112	

network) resources, the associated ports need to be opened for communication from the outside world.

■ ■ ■ ▬▬▬▬▬▬▬▬▬▬▬▬▬▬▬▬▬▬▬▬▬▬▬▬▬▬▬▬▬▬▬▬▬▬

Decimal to Binary Conversion

To convert a decimal number to a binary number (or vice versa) in Windows 2000, XP, or .NET, click on the START menu, and then click PROGRAMS > ACCESSORIES > CALCULATOR. Once you've opened the calculator, go to the VIEW menu, and choose SCIENTIFIC. The calculator will expand and give you the options for adding decimal (DEC) or binary (BIN) numbers. If you type in a decimal, such as 192.168, press the F8 key to change it to the binary equivalent. If you have BIN chosen on the menu and you type 1000010 as a binary number, press the F6 key and it will convert the binary number to its decimal equivalent: 66.

The explosive growth of the Internet is gobbling up the 32-bit IP addresses (4 billion of them) because today's IP version 4 (IPv4) originated 20 years ago and few people at the time could imagine needing more than 4 billion addresses. The new IP version 6 (IPv6) will expand the size of the IP address to 128 bits, which will bring it up to 340,232, 366,920,938,463,463,374,607,431,768,211,456 (340 duodecillion, or 34 trillion trillion trillion, or 3.4×10^{38}). That should cover us for a while.

▬▬▬▬▬▬▬▬▬▬▬▬▬▬▬▬▬▬▬▬▬▬▬▬▬▬▬▬▬▬▬▬▬▬ ■ ■ ■

Subnet Mask

A subnet mask determines which parts of the IP address are network and host identifiers. This is a 32-bit number that distinguishes each octet in the IP address. For example, as depicted in Table 4-9, 255.255.0.0 is a standard Class B subnet mask since the first two bytes are all ones (network) and the last two bytes are all zeros (host).

Keeping in mind the unique class identifiers within the first byte of the IP address (see Table 4-9), in a subnetted network the network portion can be extended to 255.255.255.0, which would subnet a Class B address space using its third byte. The first two octets of an IP address would identify the Class B network, the next octet would identify the subnet within that network, and the final byte would select an individual host. Subnet masks are on a bit-by-bit basis; thus a subnet mask like 255.255.240.0 (4 bits of subnet; 12 bits of host) can also be used.

Table 4-9 IP Address Class Identifiers

Class	IP Address	Network ID	Host ID
A	a.b.c.d	a	b.c.d
B	a.b.c.d	a.b	c.d
C	a.b.c.d	a.b.c	D

Bandwidth

Obviously, when planning to stream megadata of video through the network, bandwidth becomes exceedingly important, especially if sharing that bandwidth with other internal network resources (e.g., printers, Internet access, file servers, etc.). Depending on a number of factors, video can hog bandwidth and bring the network down to a crawl.

Typically, network devices such as routers and/or switches become the network bottlenecks. For example, if the network was designed around gigabit devices but somewhere within the building there's that old 10BaseT switch, that alone can bottleneck the entire network (especially if it's in front of the gateway). When you're engaged in the requirements gathering and surveys, even if detailed topology diagrams were provided, it's best to validate all the data and devices as part of this process. Somewhere along the line, after those diagrams were created, devices changed and that needs to be documented prior to final design and implementation of the DVS system. Otherwise, that existing device, which was originally your client's problem, will now become your problem in the form of a bottleneck.

VLAN

VLANs allow the creation of different logical and physical networks, whereas IP subnetting simply allows the creation of logical networks through the same physical network. A simple switch can create a single physical network. Different logical networks are created within the network by simply assigning different IP networks (e.g., 192.168.0.0/24, 192.168.1.0/24, etc.). As subnets are just subdivisions of a single Class A, B, or C network, they're all sharing the same physical backbone (the switch) the network broadcasts travel.

The VLAN is connected logically rather than physically, either using a different port on the physical switch or via physical addresses (MAC address). Broadcasts between these logical connections won't be seen on any other port. A VLAN is only required when a network is so large and has so much traffic that it's overwhelmed with broadcast traffic. The VLAN also allows (over a subnetted network) for one device to be connected to a switch while another device is connected to a different switch in an entirely different physical location, yet those devices can still be on the same VLAN (broadcast domain).

You need to consider using a VLAN in any of the following situations:

1. More than 200 devices on the network.
2. Heavy broadcast traffic on the network.
3. Groups of users need more security.
4. Groups of users require the same application and need to be on the same broadcast domain.
5. Multiple virtual switches from a single switch need to be created.

Video Networking

Digital video is a heavy load for any network. Compression technologies have shrunk the bandwidth requirements for video while the bandwidth pipe has increased, which has opened up more space for more video. It's the network equivalent of moving into your first home and thinking "we'll never be able to fill all these rooms!" But sure enough, if the space is there, it will get used.

The link layer and network layer protocols, which are the foundation of the TCP/IP suite, are primarily concerned with transferring data through networks. They're used by the protocol stack itself, but they aren't used directly by applications that run over TCP/IP. The two protocols used by applications are UDP and TCP. Most digital streaming video is delivered via TCP or UDP.

Transmission Control Protocol

TCP isn't the best choice for any digital video stream, especially live footage, as it requires a connection-based transmission channel. A constant connection must be acknowledged before data transfer occurs. When you're streaming live video the demands on the network increase, causing the bandwidth to close up and the acknowledgment packets to be lost. This then causes network latency issues. TCP handles the process of transmission by breaking down large segments of data into smaller packets based on the physical network used, thus ensuring that data are received at the other end before transmitting. UDP, on the other hand, is a connectionless protocol and doesn't guarantee the delivery of data sent, thus leaving the whole control mechanism and error-checking functions to the application.

In general, TCP is used when reliable communication is preferred. Many pan-tilt-zoom (PTZ) IP cameras allow TCP for the PTZ controls, while video is sent via UDP. TCP's reliability through retransmission may introduce significant delays, but doesn't fail. When attempting to reach a Web site or retrieve e-mail it's not the TCP that fails, it's usually the application or hardware. When TCP packets do arrive, they arrive without errors because the sender keeps an extra copy and waits for an acknowledgment of receipt. Once receipt is acknowledged, the extra copy is discarded. If delivery fails, the packets are re-sent until they get through to their final destination. TCP is a synchronous transmission method: if it can't sense a connection, it will continue to re-send the data until it synchronizes.

User Datagram Protocol

UDP is asynchronous with no dialog expected between the sender and receiver, which is why UDP is a better choice for live video and is very efficient in broadband environments. If you have digitized video and are delivering it as video frame 1, 2, 3, 4, 5, and so on, up to 30 fps, sudden network latency (such as you might experience with TCP) can cause a couple of frames to drop (say, frames 24 and 27). Once these frames are retrieved, they can't be inserted into their proper numeric order and obviously displaying them after frame 30 is confusing and unacceptable. Digital video is normally sent via UDP for this very reason. However, most firewalls block all UDP packets for security reasons, making it impossible for streaming video to reach desktops (more on this in Chapter 9). UDP is a very simple protocol and like IP it's an unreliable, connectionless protocol. There's no acknowledgment that any UDP data (a datagram) are ever received, but dropping a frame or two every second seems a worthwhile risk to ensure that the live video continues to stream.

UDP adds four 16-bit header fields (8 bytes) to the data sent. These fields include:

[Length Field] [Checksum Field] [Source Port Number] [Destination Port Number]

Port number, in this context, represents a transport to the application layer software port, not a hardware port. This is a virtual port that allows select protocols and applications exclusive access into the TCP and UDP networking environment. There's no physical port to plug into as this is all done through software.

Port Numbers

Both UDP and TCP use the concept of port numbers. The port numbers (see Table 4-10) identify the protocol used to send and receive the data. Most protocols have standard ports generally used for this. The use of standardized software port numbers makes client/server communications faster. The concept works like this: each compressed frame is divided into transport units and packets, first encapsulated as a transport packet (RTP, HTTP, or MPEG-2 transport), then given a UDP or TCP header, and then an IP header, such that a video IP packet transmitted over an IP network may look like the following:

[IP header] [UDP or TCP header] [RTP header] [Video payload]

The port number and the protocol field are inside the IP header for the assigned port numbers. IP uses the protocol field to determine whether the data are passed through UDP or TCP using the port number utilized by the application layer protocol. All TCP/UDP ports below 1024 are reserved and shouldn't be used, not even for serial port, video, or audio communication. The maximum value is 65,535.

Networked Video Delivery Methods

Unicast

Unicasting is a method of delivery, not a protocol. Unicast packets are encapsulated and delivered using TCP via the transport layer as most LAN and Internet traffic is in Unicast. When a source and the recipient communicate on a single point-to-point basis, data packets are solely addressed to a single recipient. No other hosts on the network process or receive this information, as they would receive their own copy of the data if

Table 4-10 The Most Common Ports* Along with the Select Protocols and Correlating Ports for DVS Application

Service	TCP	UDP	Notes
Telnet	23	23	Command line interface (handy for troubleshooting)
FTP	20	20	File transfer protocol (data)
FTP	21	21	File transfer protocol (control)
SSH	22		Secure shell
HTTP	80		Hypertext transfer protocol (e.g., for Web browsing)
MS NetMeeting	1024, 1503	1024	Video conferencing
HTTPs	443		Secure HTTP (SSL)
QuickTime 4	RTSP	RTP-QT4	Streaming audio, video
RTSP	554		Real-time streaming protocol currently described in RFC 2326
SOCKS	1080		Internet proxy
PPTP	1723		Virtual private network (VPN)
MS NetShow	1755	1755, 1024-5000	Streaming video
MADCAP (multicast address dynamic client allocation protocol)	2535	2535	Multicast address dynamic client allocation protocol (MADCAP). Defined in RFC 2730
Windows remote desktop protocol (RDP)	3389		Terminal server
Yahoo Messenger – Webcams	5100		Video
AIM Video IM	1024-5000	1024-5000	Video chat
Multicast DNS	5353	5353	The regular DNS port is 53
pcAnywhere	5631	5632	Remote control
eShare Chat Server	5760		
eShare Web Tour	5761		
eShare Admin Server	5764		
VNC	5800+, 5900+		Remote control
RTP-QT4		6970-6999	Real-time transport protocol (these ports are specifically for the Apple QT4 version)
VDO Live	7000	User-specified	Streaming video
Real Audio and Video	RTSP, 7070	6970-7170	Streaming audio and video
Common HTTP	8000, 8001, 8080		Alternatives for Port 80
RTP-iChat		16,384-16,403	Used by Apple iChat AV
RTP		16,384-32,767	Real-time transport protocol in general is described in RFC 3550

*You can learn more about TCP and UDP ports at http://www.iana.org/assignments/port-numbers.

they were the sender. Surfing the Web is done via Unicast – each visitor's interaction on the Web is considered point-to-point communication. The person surfing the Web caches a copy of the Web pages visited onto their temporary Internet files folder, thus creating the process of source and recipient sending and receiving information.

When a digital video stream is transmitted via Unicast, it streams a single (uni) stream of video per host computer. Each host receives its own unique broadcast of the streaming media. If that stream is an average of 1.5 Mbps, and six clients decide to view it concurrently, that's 9-Mbps streaming throughout the network (see Figure 4-10). Unicast follows traditional TCP/IP Class A, B, or C addressing attributes by delivering the single message to a single unique address, closing that loop between one IP address and the other.

Unicast eats up network bandwidth as the demand for more streams creates more streams traveling through the network. Depending on the initial requirements for the system, the multiple streams may go unnoticed, and there may be no network latency for years until the demand outgrows the network bandwidth, which could bring the network down to a crawl.

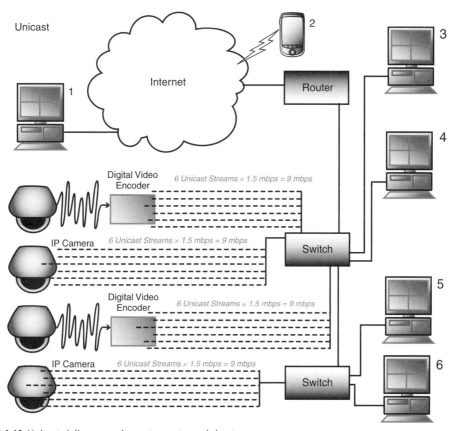

FIGURE 4-10 Unicast delivers a unique stream to each host.

Anycast (Broadcast)

Anycast, or broadcasting, is similar to television broadcasting because it delivers data from one single source to everybody on the network, whether they want to receive it or not. Unlike television broadcasts, using broadcasting for video can cause network storms, so it's not recommended for IP networking (see section Broadcast Storms).

Multicast

Multicasting is defined as a single source sending to multiple recipients on a network when the receiver broadcasts a signal for acceptance. Multicast has its own Class D IP addressing scheme controlled and assigned by the Internet Assigned Numbers Authority (IANA). This means that all IP multicasts are in the range of 224.0.0.0-239.255.255.255. This unique IP address range is used only for the destination address of IP multicast traffic as the server, encoder, or IP camera delivering the multicast datagrams is always the Unicast source IP address.

Multicast dramatically reduces network traffic by delivering a single video stream to multiple receivers (see Figure 4-11). For example, by using Unicasting in a DVS system

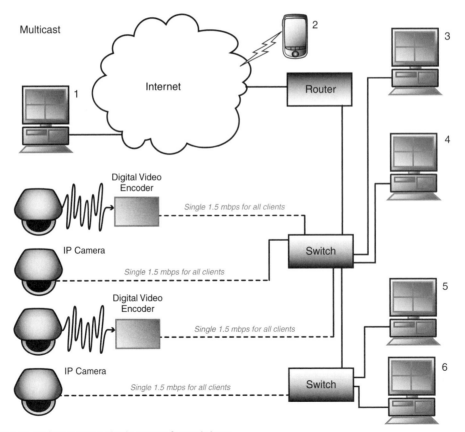

FIGURE 4-11 Multicast uses a single stream for each host.

with redundancy archivers there's a video stream per backup archiver, plus (depending on the VMS used) a single video stream per client station. Thus, with three backup archive servers and three client stations, one camera will send six different streams. Multicasting, on the other hand, was created for audio and video transmission over IP networks, allowing a single video stream to be delivered to multiple recipients. Multicasting is also a one-way connection so it can't be transported via TCP, which requires confirmation of connectivity. Multicast, by its very nature, is only sent via UDP (so make sure those UDP ports are open).

Understanding Broadcast and Multicast Packets

Broadcast packets are sent to all of the devices on the network, while multicast packets are sent to a group of devices on the network if configured as such (otherwise, without spanning tree configuration or Internet Group Management protocol [IGMP], multicast packets are broadcast to all devices as well). Unicast packets are sent from one source device to a single destination device using IP addresses. Multicast and broadcast packets are identified by a unique numeric signature in the first byte of the destination media access control (MAC) address (see Figure 4-12).

Excessive broadcast and multicast packets are detrimental to network performance because all of the devices must process each packet and the additional processing overhead may cause denial of services or drop incoming packets. It's always best to control the traffic with whatever tools are available. This is where multicast groups, IGMP, and spanning tree can control the traffic and avoid broadcasting storms.

Broadcast Storms

Broadcast packets and multicast packets are a normal part of your network's operation. When broadcast and multicast traffic is abnormally high for your network it may be due to a broadcast storm. The best way to identify the problem is by analyzing network response times and network operations. As a broadcast storm progresses, users can't log in to servers, encoders, or cameras, and as the storm worsens, the network becomes unusable. Storms occur when network equipment is faulty or configured incorrectly, when the spanning tree protocol isn't implemented correctly, or if poorly designed programs are used to generate broadcast or multicast traffic.

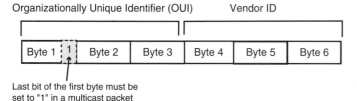

Organizationally Unique Identifier (OUI)   Vendor ID

| Byte 1 | 1 | Byte 2 | Byte 3 | Byte 4 | Byte 5 | Byte 6 |

Last bit of the first byte must be
set to "1" in a multicast packet

FIGURE 4-12 Anatomy of a multicast packet.

Multicast Group

Multicast is based on the concept of a group interested in receiving the same data stream. This group doesn't have to have any physical or geographical boundaries as long as they have access to the designated LAN or WAN. Multicast groups are created to reduce the broadcast traffic generated by the host signaling to receive the multicast stream. This is done using IGMP snooping. If it's not done then the very nature of multicast broadcasts turns them into what I refer to as "multicast noise." Multicast noise is a broadcast storm to every active Layer 2 switch port on the network, whether requested or not. This may not be an issue within a bandwidth-rich network or for anything other than streaming live video streams, but when the pipe shrinks or is limited, based on wireless signals, interference, or older switches, the noise becomes a problem.

Real-Time Transport Protocol

Real-time transport protocol (RTP) is a thin protocol typically sent via UDP. It doesn't actually guarantee real time, but it does enhance the control and synchronization streaming media. MPEG-1 and MPEG-2 provide their own synchronization for video conferencing (MPEG-1 system stream and MPEG-2 transport stream to name a couple), but usually demand too much bandwidth for painless streaming of multiple digital video cameras. Although designed for online streaming, where the bandwidth can be limited and the network unpredictable, MPEG-4 doesn't have its own audio/video synchronization. RTP provides the required time stamps for audio/video synchronization for MPEG-4. Many IP cameras and encoders provide interoperability with other devices and support RTP for transport (Figure 4-13).

Real-Time Streaming Protocol

Real-time streaming protocol (RTSP) is a control protocol for RTP. You wouldn't find RTSP without RTP, as RTSP calls the Unicast stream (doesn't work with multicast) into the VMS and/or video player interface (see Figure 4-14).

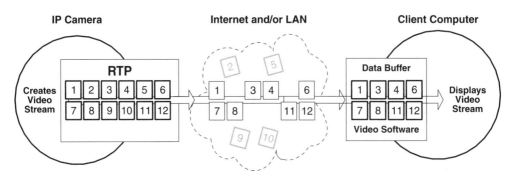

FIGURE 4-13 How RTP works.

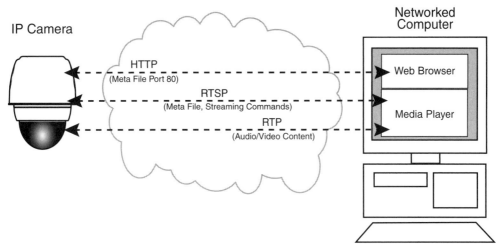

FIGURE 4-14 RTP communication between a host computer and an IP camera.

Hypertext Transfer Protocol

The hypertext transfer protocol (HTTP) is most often confused with hypertext markup language (HTML), which is the basic coding language of a Web page. It's important to recognize HTTP because most corporate firewalls allow the default HTTP port 80 to pass traffic, thus making it possible to use a Web browser to access the Web server interface in today's IP cameras and digital encoders by using a software plug-in to view the cameras or to use a plug-in within a VMS system from anywhere there's Internet access.

Remote Access – Your Home away from Home

The definition of remote access as it relates to this book is accessibility to the new security cameras and/or VMS from anywhere at any time. The core requirement to perform this magic of technology is online access. Internet access provides the link needed to the server hosting the VMS and/or the built-in interface available in today's IP cameras and digital encoders. The real trick is gaining access without network vulnerability by opening up the network to the outside world. This can be done as detailed in Chapter 9.

Lessons Learned

Before you even begin to attempt to uncover the mystery of your DVS system, you need the tools. These tools are a mixture of hardware and software tools and cover a multidimensional view of troubleshooting complex systems. As Figure 4-15 depicts, a complex system uses mechanical, electrical, communications, and software components as well as modules concurrently and in relation to each other to function properly.

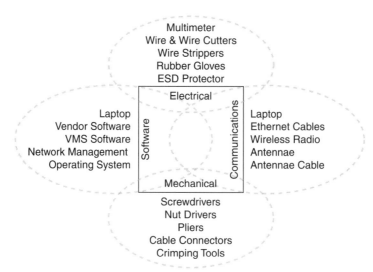

FIGURE 4-15 Multidimensional troubleshooting.

If a camera is down, meaning there's no video appearing within the VMS software, the problem could be anything from the camera PTZ motor to the lack of power to the camera, an encoder, a network switch, etc.; it could be a completely unrelated network problem, or a software or firmware problem (in Chapter 5 another wireless level is added).

What I hope to offer is some experience that may help save time when troubleshooting a DVS network.

Troubleshooting Laptop

Don't assume that since you're out in the field attempting to check on a camera that there wouldn't be a need for a service troubleshooting laptop. This special laptop is void of select VPN applications and resource-hogging collaborative (and virus-introducing) applications such as Lotus Notes or Outlook. There shouldn't be any video games or DVD players or anything that may conflict with the applications required to install, configure, troubleshoot, and maintain the DVS system. It needs to be a clean system without any firewall software and the bare minimum of antivirus software (since it will rarely see anything other than the DVS network, it should be safe).

The other option is a virtual machine (VM). A VM is a computer and OS completely separate from the host computer and OS. Microsoft offers a free, exceptional VM software called Virtual PC. Just install it on the select laptop (make sure there's plenty of hard drive space and memory to share) and then once Virtual PC creates a separate "virtual" computer on your computer, you'll need to install a new OS. This is a great way to create a "virgin" machine free of any applications that may conflict with the required DVS software tools and applications.

Software Troubleshooting Tools

There are a few applications built into Microsoft Windows that are very useful when troubleshooting DVS equipment: file transfer protocol (FTP), Telnet, and administrative devices such as PING, NETSTAT, and TRACERT.

PING is a simple utility that's extremely valuable when troubleshooting network communications. It sends a signal from one computer to another and returns it, confirming a clear connection. NETSTAT offers data on protocols, ports, and packets, and TRACERT traces the actual path of a packet through the network. These tools are used at the command prompt, which can be accessed in Windows as follows:

START > RUN > type CMD and press ENTER

The following commands are useful tools when troubleshooting video networks.

PING

At the command prompt, type PING and then the desired IP address that you wish to communicate with and then press ENTER. You'll either receive a Reply from, a Request timed out, or a Hardware error message. Either way, this simple test can help you choose the next troubleshooting step.

IPCONFIG/ALL

When you type IPCONFIG/ALL at the command prompt, and then press ENTER, it will display the IP address information for each one of the network interfaces and their physical MAC addresses.

NETSTAT

Typing NETSTAT at the command prompt displays all active connections to the computer. This comes in handy when attempting to determine if a camera or encoder is connected to a VMS server. There are also additional functions:

NETSTAT-R displays the routing table.
NETSTAT-A displays all connections and listening TCP/UDP ports.
NETSTAT-E displays Ethernet statistics and can be combined with the NETSTAT-S option.
NETSTAT-S displays per protocol statistics.
NETSTAT-N displays addresses and port numbers in numerical form.

Other third-party software applications that may come in handy include Net World Scanner, Ping Scanner Pro, TCP Net View, and FastResolver.

What Usually Goes Wrong

There's a complexity to a digital video network that goes beyond a typical enterprise network. Emergencies may include connectivity to the e-mail server or even a CCTV implementation, which is a closed, isolated system running on coax. This section covers some of the problems previously encountered and the solutions or workarounds used, but this isn't the one-size-fits-all solution to all problems. There are far too may variables and different design architectures to be able to immediately determine the cause of every problem. It takes time and patience, but I'm happy to say that it's rarely anything so obscure that it eludes you for long.

No Video

The most immediately recognized problem in a DVS system is when a "camera" is down. This can be presented as a black screen, a split-colored screen, a Waiting for message, or a small icon of a camera with an "X" over it. What's displayed depends entirely on the VMS used, the camera, and the design architecture. For example, when using an analog camera connected to a digital video encoder, there's another layer of complexity to the straightforward IP camera, which has the encoder built in. The cables from the analog camera to the digital video encoder may be the cause of a problem and the power to the camera and/or encoder may be out. This may be due to the surge protector or a bad power inverter. There's also the Ethernet cable from the encoder into the network, the coax cable from the camera to the encoder, etc., to be concerned about. See troubleshooting flowcharts in Chapter 3 for more information.

Video Networking Design

Good design practices include avoiding Layer 2 switching in any video surveillance networking environment, and especially "stupid switches." These "wannabe" hubs have no smarts and no functionality (spanning tree protocol or IGMP) other than passing data and no method of manageability. In general, Layer 3 switches are more costly, especially if Layer 2 switching domains are all that's needed initially. Once the requirements change and more functionality is needed (e.g., digital video streaming), problems rear their ugly head.

As explained in the Multicast Group section, flat Layer 2 networks are created through lack of central control and management, lack of planning, and rapid growth, but mostly because they're cheap. Unfortunately, funding or political support for a good network topology is nonexistent; they just have to be good enough (beware the "good enough" mentality as it tends to come back to haunt you). Implementation of Layer 2 switch topology always looks attractive to the bottom line because you can plug them in and they just work. Switches pass data efficiently.

For more information about good design practices for switched networks, see www .cisco.com, which is full of tutorials, white papers, and helpful hints.

Network Security

As mentioned before, the key to a secure network is controlling traffic. The open TCP and/or UDP ports must be for a specific purpose; otherwise they should all be blocked, closing the door on malicious intrusion. How important is this bit of advice? If the facts explained in Chapter 1 aren't a wakeup call, then consider the experiment I earlier outlined in the section Firewall.

Incompatibilities

Before determining the network topology is at fault, the first course of action for any DVS system is the hardware. As I mentioned before, switched networks simply pass data. Unless there's a router or firewall within the topology with blocked access ports, then a common problem is firmware. Sometimes the IP camera and/or video encoder isn't compatible with the latest version of the active VMS, and checking the manufacturer's Web site and VMS software Web site for updates and upgrades in both firmware for the devices and patches for the software can eliminate troubles.

IP Networking Troubleshooting

Figure 4-16 depicts the troubleshooting path when there's no connectivity between two devices on an Ethernet network. Those two devices could be a VMS workstation and VMS server, a workstation and router, an IP camera and/or encoder, and VMS server. Regardless of the two devices used, as long as they're on the TCP/IP network and each has an assigned IP address, this flowchart will help you uncover the problem.

If you have experience troubleshooting Ethernet networks, then this will seem very familiar. Because Ethernet is a mature and solid communications technology, problems are rather unusual. However, a DVS system invites additional layers and a new set of devices that may create new problems or just turn the previously insignificant ones into rampaging rhinos.

The reason for no connectivity between two devices on an Ethernet network varies based on how that network is architected and if there are any other devices between the two devices in question. Once the obvious possibilities are verified (e.g., power to all devices and secure connection into the network), the next step is to determine if there's any communication at all between the two devices and if the software drivers for the network adapters are up-to-date.

Using software tools built into Windows, such as PING and PATHPING, provides an initial diagnosis of the problem. If an IP device (e.g., computer, IP camera, digital video encoder) is inaccessible through protocols such as HTTP, FTP, or through a client/server application, the PING tool (using the ICMP protocol data unit) can still "see" the device (unless the ICMP protocol is blocked within the firewall configuration).

A reply from the device using PING confirms that the device is indeed on the network and the problem is something other than a physical disconnection. This is when you begin exploring blocked ports and protocols. For example, if attempting to access an IP camera with its built-in Web server, the protocol is HTTP, which uses port 80. If port 80 is blocked, accessing that device using HTTP is impossible unless within the firewall

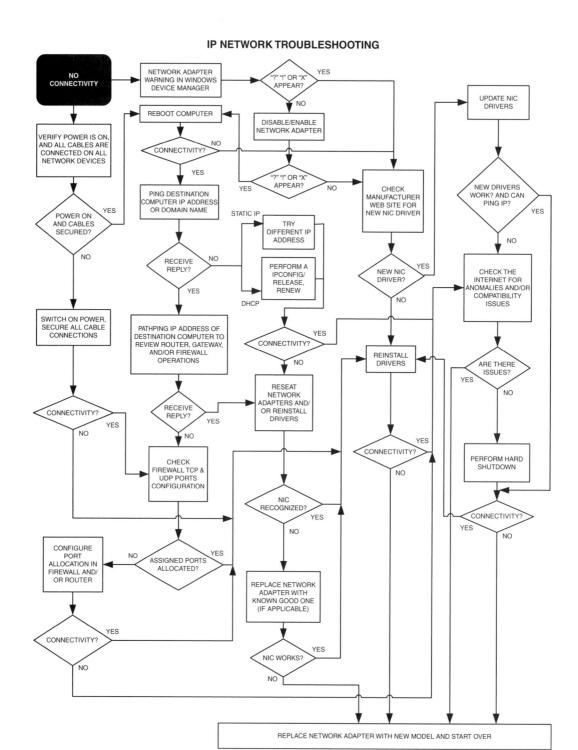

FIGURE 4-16 Network troubleshooting flowchart.

the HTTP port was changed to another port (8080 is also a popular choice). Check Table 4-10 to understand the pathway to and from the device using the various protocols offered by the IP camera digital video encoder manufacturer. Most IP cameras and the latest model digital video encoders provide a Web server built in to offer HTTP access for live viewing and configuration of the unit. A few manufacturers, such as Axis, also provide an FTP method of access and configuration, either using the Windows built-in FTP client by opening Windows Explorer and typing ftp://ipaddress or their own software application (Axis Camera Management).

Other manufacturers use a proprietary communication protocol that doesn't rely on IP addressing, but instead utilizes unique parameters and still uses TCP/UDP or multicast ports to communicate. For example, Verint digital video encoders use their own proprietary VSIP protocol, which by default uses port 5510. However, if the digital video encoder isn't set to port 5510 or if port 5510 is blocked by the firewall in the network, then their configuration tool (SConfigurator) won't see it. SConfigurator has a Discovery mode that reaches out into the network to recognize any Verint devices. SConfigurator can broadcast that signal using the broadcast address of 255.255.255.255 or using multicast, through the multicast address of 224.16.32.1. The VSIP port 9541 is considered the "common" VSIP port that recognizes all Verint devices on the network, no matter the VSIP configuration on the individual units.

Another possibility is a corrupted device driver, which an uninstall and reinstall can usually rectify. Finally, always check the manufacturer's Web site for updates and patches, not only for the DVS devices, but also the VMS server and workstation.

Drivers

Drivers are the software APIs that allow the OS to communicate effectively with additional hardware components. The OS could be the version of Windows or Linux used on the VMS server or the firmware that "drives" the IP camera and/or video encoder. As mentioned in Chapter 3, many IP cameras and video encoders use an embedded version of Linux as an OS, and Apache (an open source Web server) as the HTML interface. When these devices require communication and/or control through another device, a software component is required, whether it's already built into the VMS software or is a required download. For example, integration of an existing analog PTZ camera to a new encoder may require a driver for that encoder to understand the analog camera's "language." Otherwise, the PTZ won't work or will have limited functionality.

Windows Update

When using a Windows OS, the Windows Update feature can be a godsend or the messenger of doom. Unfortunately, it's a necessary evil, especially if the VMS server is connected to the Internet. The Windows OS is 90% of the marketplace, making it the most attractive target for relentless hackers and crackers. Windows Update provides an automated method of plugging the holes discovered during normal everyday operations when there are hundreds of hackers working with computer farms all over the world trying to uncover new vulnerabilities in 95% of all computers worldwide, so they can hijack, infect, or steal pertinent information for their own malicious agenda. The value of a dominant OS for expanding the usefulness of the PC is also its own worst nightmare. The issue with Windows Update is that its primary objective is to secure Windows and eliminate any liability that may arise from new vulnerabilities. This can and has come at the expense of patching and/or replacing software components that may suddenly become a security threat with updated versions. Sometimes these new versions don't take any add-on or third-party components into consideration, so the devices may cease to function or become erratic.

Chapter Lessons

When working with networked video, here are a few key factors to remember:

- Always discover if there's an existing detailed network topology or uncover, through due diligence, the network capabilities and limitations before adding DVS traffic.
- Maximize the use of existing networking infrastructure without jeopardizing the integrity of the existing network or the performance of the new DVS network.
- It's best to separate the DVS Ethernet traffic from network traffic either using physically separate hardware or through VLAN configurations. Stick to fiber interconnects between IDF and MDF locations, if possible.
- Always design for the maximum bandwidth possible, making way for any future growth.
- Only use network switches that provide multicasting traffic management and control to avoid multicast storms.
- Use uninterrupted power supplies (UPS) as backups on all security devices. What's the use of having video surveillance or additional digital security if cutting power will shut it all down?
- Whenever running fiber through an outdoor conduit, consider using innerduct for added protection from damage or interference.
- Keep all networking equipment within an environmentally controlled room or enclosure (avoid basements).
- Only open the specific firewall ports required to operate the VMS system to maintain the network security integrity.

5 ▪ Wireless Networked Video

Introduction

Wireless communications is part radio frequency (RF) technology, part artistry, and part magic. How technologically spoiled we've become is apparent when we don't realize that having a cellular teleconference among three people, each driving down an expressway on a different continent, is nothing short of magic. Just ask anyone who worked on the technology to make it all happen. When working with wireless communications, you're entering the RF world of waveforms once again, but these waveforms, open to the outside world, can succumb to a number of different problems including interference, noise, weather, and vibrations, not to mention the limitations of the chosen software and hardware. You may see it as a "simple" solution when there's no wired alternative, but it's not an "easy" solution. However, when it works, it is indeed magic.

Introduction to RF

The RF spectrum is divided into small chunks for a huge number of applications such as AM and FM radio, television, cellular networks, walkie-talkies, satellite communications, military applications, and even to send and receive signals into outer space. Transmitters and receivers are modulated to "hear" only the specific frequency programmed, but depending on the power and frequency, there may be "bleeding," which can cause interference. There are also software and hardware limitations and programmability issues. It's not a simple task to create the magic RF box, especially for the implementation of a video network using the available standards. It would be much easier if there were a standard, exclusive frequency set aside specifically for video networking (as with television), but RF shares bandwidth and channels with e-mail, Internet access, and backhaul connections between offices and buildings. Thus, we see scenarios where five network television stations are broadcasting in urban areas, each with their own exclusive piece of the spectrum, while there may be a hundred individuals trying to share a minute portion of the unlicensed spectrum for wireless networking.

Without Wires?

Wireless doesn't mean "without wires" unless the design specifications require a closed wireless local area network (WLAN) between wireless client workstations or laptops and the individual wireless IP security cameras all linked via a wireless access point (AP). At some point you'll be connected to a wired backhaul if you want remote access (from anywhere else) via the Internet. As discussed in Chapter 4, many resources are shared within a local area network (LAN) such as printers, file servers, storage, and the Internet, so without wires the accessibility to the system is limited.

If you're using digital video encoders and IP cameras, there's also the Ethernet link between the actual encoder and/or IP camera and the wireless radio to consider as well as the power link. Wireless networked video provides another option for data transmission, but can't replace the basic need for power. So keep in mind that if there's a new conduit design for power, the cost isn't much more to have parallel conduit runs – one for power and the other for data. Wireless is a substitute under the following conditions:

1. There's power at a desired location, but no data access.
2. The distance from the camera to the closest data port exceeds the cabling requirements.
3. There's a cluster of cameras that could benefit from wireless mesh networking redundancy.

This is where due diligence becomes essential. I can't tell you how many potential data pathways I've uncovered through insight and perseverance. Sometimes the floor plans and diagrams aren't enough and getting your hands dirty by opening hand-holes and lifting manhole covers and crawling through attics and crawl spaces reveals those data pathway possibilities. Sheer determination and commitment to the success of the project are essential. Wireless does work and works well, but unlike an insulated and protected copper wire or fiber cable, wireless is RF technology that can be affected by outside influences such as interference from other radios, microwave ovens, and even the weather. Throughout my experience, even though I'm considered a wireless video expert, I've come to the conclusion that wired is always better than wireless and that my expertise comes from a willingness to perform the extra work required to make those "trouble spots" actually work and work well.

Radio Frequency

RFs are high-frequency alternating current (AC) electromagnetic signals that travel through a copper wire and radiate from an antenna through the air as both transmitter and receiver. These AC signals, once in the air, turn into radio wave or waveforms. How those radio waves travel is based on the type of antenna (see the Antennas section).

An RF amplifier and/or a high-powered antenna is used to increase the RF gain or amplitude. RF loss occurs when there's resistance in the signal such as traveling through RF cable and connectors or impedance with mismatched cables and connectors.

■ ■ ■ ━━━

Hertz and Wavelength

An RF wavelength is measured in kilometers, centimeters, or millimeters. It's the distance between two points to complete one cycle at a particular frequency:

1 Hertz (Hz) = One cycle (wavelength) per second
Wavelength = Distance between two adjacent corresponding locations on the wave train

The measurement in Hertz is the period cycle and 1 Hz cycles a wavelength in one second, whereas 1 GHz (1,000,000,000 Hz) cycles 1 billion cycles per second; the lower the frequency, the larger the wavelength.

━━━ ■ ■ ■

Measuring for RF loss assigns a sensitivity threshold for each band, which defines the point where the radio can distinguish between a signal and just background noise. The key is to tune in to the right channel and get the clearest signal possible to clearly understand the message, similar to tuning an FM radio to a specific channel. The farther away from the transmitter, the more difficult it is to get a clear signal because the distance opens up additional interference to degrade the signal.

RF reflection can increase the signal strength by adding the reflected signal onto the initial signal, but that isn't necessarily a good thing. This phenomenon is called "multipath" and can degrade the main signal and even cause holes in the area of coverage. Multipath is caused by any number of surfaces, depending on the frequency. Even though the lower frequencies with larger waveforms can better travel through solid objects, they will be immediately dampened. They can still be reflected by metal and water, which causes multipath interference. Even rain can affect an RF signal. Although the frequencies more severely affected are above 11 GHz, rain can still reduce bandwidth.

The Federal Communications Commission (FCC) regulates and monitors the use of RF electromagnetic energy flowing through the atmosphere. There are both unlicensed and licensed bands that can be used. Although manufacturers of radio equipment must follow the strict FCC regulations, they don't monitor who uses those RF frequencies.

The piece of the RF pie designated for WLAN applications falls within the microwave band (see Figure 5-1), which starts at 1 GHz and goes up to 300 GHz. WLAN uses the 2.4-GHz ISM band, which is license-free and used by microwave ovens, cordless phones, and Bluetooth. There's also 802.11b (Wi-Fi) WLAN, 4.9 GHz for public safety, 5.5-5.7 GHz for 802.11a WLAN, 5.725-5.85 GHz ISM U-NII band (license-free), WLAN 802.11a at 54 Mbps, and 108 Mbps plus for 802.11n multiple-input multiple-output (MIMO) (more in the section MIMO later in this chapter). The 700- and 900-MHz bands are relocated for public safety, but require licenses for use.

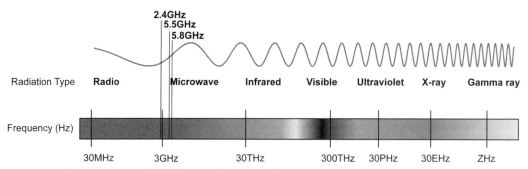

FIGURE 5-1 The RF spectrum and the location of the WLAN frequencies.

Our Piece of the RF Pie

You can download the entire FCC regulations RF chart from www.ntia.doc.gov/osmhome/allochrt.pdf.

There are other types of signal distortion and effects that may change how the RF signal behaves, but for the purposes of this book they will be defined as needed and any other information you may need can be researched at www.ieee.org or more detailed RF textbooks.

RF signals can be distorted by refraction, diffraction, scattering (when bouncing off the waves of turbulent waters), and absorption, all based on surface and/or atmosphere substance. Voltage standing wave ratio (VSWR) is a man-made effect caused by using mismatched impedance (resistance to current flow measured in ohms) between devices in an RF system, which can unknowingly degrade the signal power and quality. The only time this appears to be a problem is when the same configuration and installation are done at another location and the amplitude is significantly higher or lower than at the previous location. There's nothing like experience to force you to check the equipment procured and shipped on location for implementation (especially if working in subzero weather).

The wireless 802.11 standard specifies the physical sublayer, media access control (MAC) sublayer, and MAC address management in two basic service sets: ad hoc and infrastructure. The ad hoc service is known as the Independent Basic Service Set (IBSS), while an infrastructure service set is managed by an AP.

Access Point

An AP is a switch (Layer 2 Bridge) with two networking technologies: IEEE 802.3 Ethernet on one side and IEEE 802.11 wireless on the other side. An AP is what provides wireless connectivity to distributed resources such as a LAN, with printers and storage,

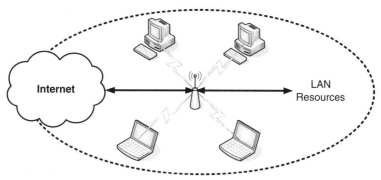

FIGURE 5-2 Wireless AP infrastructure topology.

and/or the Internet. The AP can only provide wireless access to those clients and/or stations securely associated with its encryption and authentication. Those clients/ stations are oblivious to each other (unless designed to share resources on the network), and can only communicate with the AP creating an access point/station (AP/STA) mapping based on accessibility (see Figure 5-2).

Basic Service Set

A Basic Service Set (BSS) forms an ad hoc self-contained network with station-to-station traffic flowing directly, receiving data transmitted by another station, and only filtering traffic based on the MAC address of the receiver (see Figure 5-3).

Extended Service Set

The Extended Service Set (ESS) consists of one or more interconnected WLANs integrated into LANs that appear as a single BSS to the logical link control layer. Any client/station can disassociate from one AP and associate to another AP, depending on traffic thresholds and signal strength.

Service Set Identifier

The Service Set Identifier (SSID) is the WLAN "network name." Similar to how a wired computer must associate itself with a specific work group or active directory domain, the wireless client station must associate itself with the SSID. Each network uses an SSID, a 32-octet string to separate one network from another for bandwidth, authentication, and security reasons. This limits

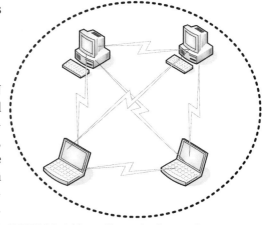

FIGURE 5-3 Ad hoc self-contained network.

client stations to associating only with APs with matching SSIDs, unless the client station is configured for SSID = ANY or the closed system feature is turned off. Although this isn't part of the standard, it's available on most commercial APs and isn't recommended for security reasons. It's best to maintain control of who and what can sniff your network. Turning off the broadcasting of the SSID, so that no other client stations can see it, is a good idea.

Beacons

Beacons are periodically sent from AP-to-client stations (in an infrastructure mode) or station-to-station (ad hoc mode) to synchronize the communication between associated members. Beacons contain:

- Time synchronization information (including beacon interval)
- Channel information
- SSID
- Traffic indication map (TIM)

Beacons manage client stations in a coverage area of multiple APs with the same configuration criteria, assisting in expediting associations when roaming between APs.

Hidden Node

The hidden node issue is an illusive wireless nuance that can be quite frustrating if there's no way to move the physical location of each wireless device or tweak the request-to-send/clear-to-send (RTS/CTS) function (see the next section Request-to-Send/Clear-to-Send) to compensate for the jamming problem. The hidden node dilemma happens primarily in wireless ad hoc networks where there are multiple nodes (client workstations) attempting to communicate with other nodes concurrently. The hidden node is ignored conversely, as the other nodes are too busy talking to each other to notice. For example, node A sends a signal to node B, but node C doesn't detect it, so node C might also start sending to node B to try to get its attention. This creates a collision of messages at node B, corrupting and losing both messages. This happens most often in larger ad hoc networks and wireless mesh networks, where the problem can affect the performance of the entire mesh.

Request-to-Send/Clear-to-Send

Part of the 802.11 standard and a solution (although some radios handle this better than others) to the hidden node issue is a MAC address level RTS/CTS, which adds a bit of overhead but avoids latency and dropped nodes.

The exchange of RTS/CTS data prior to the actual video frames is one means of managing all the nodes yelling at each other to get everyone's attention. A node (AP, client station, or mesh radio) receives the RTS and responds with a CTS frame. The node must receive a CTS frame, which contains a time value that alerts other stations to hold off from accessing the targeted node, while the node initiates the RTS transmission.

This RTS/CTS handshake provides control over how each node communicates with each another, keeping the signal active for streaming video. The main reason for implementing RTS/CTS is to minimize collisions among hidden nodes. If there's no hidden node problem, it's best to deactivate this function because it adds overhead and may be detrimental to a system demanding high frames per second (fps) and bit rates.

Interference

RF interference is caused by two or more radios, each on different wireless networks, using the same frequency. Interference can also be from 802.11 and non-802.11 devices, microwave ovens, Bluetooth, wireless telephones, radar signals, etc. To avoid interference there are other elements that need to be discovered when doing the wireless site survey (see Chapter 6). Table 5-1 shows examples of additional RF barriers.

Interference can also be unintentional, as each manufacturer has different numerical channel assignments to select frequencies or overlapping frequencies. For example, Brand A may have the same center frequency as Brand B but on a different bandwidth frequency that overlaps the channel on Brand A. When working with other agencies and companies using the same RFs, it's best to follow the frequencies instead of the channels.

Interference can also be intentional and used to create denial-of-service attacks to bring down the WLAN. Remember, there's no control over who can use the unlicensed bands.

Direct sequence spread spectrum (DSSS) is better at resisting interference and noise, but with any RF technology, when there's interference throughput levels can drop down to zero bandwidth. A few nearby frequency hopping spread spectrum (FHSS) systems can cripple any FHSS or DSSS system, although a DSSS system continuously transmits on every frequency in the band so the FHSS systems won't be able to find a clear channel and consequently degrade.

Table 5-1 Additional RF Barriers

RF Barrier Description	RF Severity	Examples
Air	Minimal	Unless raining
Wood	Low	Partitions, wall studs
Plaster	Low	Interior walls
Synthetic material	Low	Interior walls, plastic siding
Asbestos	Low	Ceilings
Glass	Low	Windows
Water	Medium	Damp wood, aquarium
Bricks	Medium	Interior and exterior walls
Marble	Medium	Interior walls, floors, structures
Paper	Medium	Books on bookcases, files in file cabinets
Concrete, wire mesh	High	Floors, exterior walls
Chicken wire mesh	High	Bulletproof glass, security booths
Metal	High	Desks, metal partitions, reinforced concrete, automobiles, fire escapes

The FCC regulates the amount of power permitted by each band, licensed or unlicensed, but it doesn't monitor the number of transmitters a single source can use or how closely they can operate. For example, if you're using the 2.4-MHz band for video surveillance solutions, then the signal can travel through walls. However, each transmitter used by each office in the entire building, coupled with any 2.4-GHz cordless phones still in operation, would create a constituent fight for power with all transmitters playing dueling banjos. This situation would also create an inadvertent attempt to cripple the wireless band, which could be done by anyone who wants to shut down the WLAN to shut down the cameras. This is where directional antennas can help by providing a more laser-focused signal that ignores interference better. This works better than an omni-directional antenna that transmits in all directions.

Line of Sight

Line of sight (LOS) is defined as having a clear visual between the transmitter and receiver without any obstructions. Although microwaves behave similarly to light and sound waves, the higher the frequency, the more diffusion and refraction when traveling through solid objects or even rain and fog.

As with light and sound, best practice is to provide a clear LOS from transmitter to receiver with no RF noise in between. LOS becomes imperative the farther the signal must travel. Depending on the height of the transmitter and receiver, RF can travel for miles; at 11 miles the curvature of the earth becomes an obstacle.

Fresnel Zone

The Fresnel (pronounced "FRE-NELL") zone occupies an ellipsoid area between the transmitter and receiver LOS. This zone is an important element in determining the best signal strength, especially when select obstructions are visible within the LOS between the transmitter and receiver. As depicted by Figure 5-4, 70% of the very center of the LOS signal must be clear to prevent multipath interference and signal deterioration. The transmitter and receiver must be at the same height to provide optimal signal strength, unless flat panel directional antennas are used to align the signal (more on this in the later section Antennas).

The radius of the Fresnel zone, at its widest point, can be calculated using the following formula:

$$R = 43.3 \times \sqrt{d/4f},$$

where
d is the link distance in miles,
f is the frequency in GHz (e.g., 2.4 GHz, not 2400 MHz), and the answer
R is in feet.
An example would be a 2.4-GHz link at 5 miles (8.35 km), resulting in a Fresnel zone of 31.25 feet (9.52 m) at its widest point.

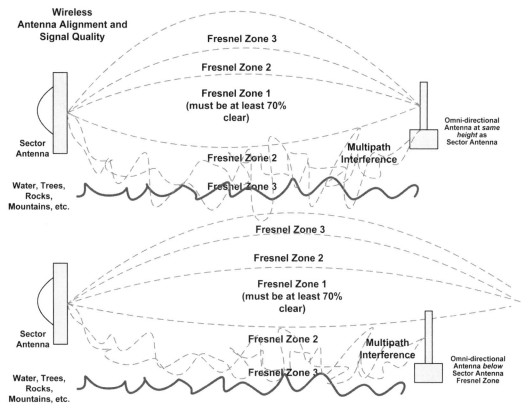

FIGURE 5-4 Fresnel zone must be 70% clear for optimal performance.

Quantification of obstacles within the Fresnel zone is important to determine the amount of disruption they will cause (see Figure 5-4). Up to 30%, or in some cases 40% depending on the type of obstruction, of the Fresnel zone can be blocked with little or no disruption. This has little bearing on the new 802.11n MIMO technology, which uses multipath interference to its advantage.

If tree growth or new construction blocks more than 20% of the Fresnel zone, then raising the transmitter and receiver usually corrects the problem. In a mobile environment, such as indoors, the RF signal bounces off and moves through many obstacles, so the Fresnel zone is constantly changing and users normally dismiss the poor signal strength as a "dead zone" or weak signal strength.

Antennas

An antenna is an electronic component made to efficiently radiate and receive radiated electromagnetic waves. An antenna element, without amplifiers and filters, is typically understood to be a passive device. There's no conditioning, amplifying, or manipulating the signal. The antenna merely creates an amplifying effect with its shape, resulting in a more narrowly focused beam. Just as a flashlight appears to be brighter and shine farther than an ordinary bulb simply because it focuses the beam in one direction, the RF wavelengths can be focused to travel farther by taking the typical 360° omni-directional dipole antenna and narrowing the beam in one direction at 30° (see Figure 5-5). This is how directional antennas work to move the wavelengths farther. The most important antenna characteristics are radiation patterns, power gain, directivity, and polarization.

FIGURE 5-5 Directional antenna transmitting video data to a portal base station.

Intentional Radiator

The FCC defines the intentional radiator as all the components from the radio to the connector on the antenna, but not including the actual antenna. FCC regulations are strict regarding output power for all RF devices. The details of these regulations can be found in *Radio Frequency Devices*, Part 47 CFR, Chapter 1, Section 15.247, October 1, 2000. You and your wireless vendor should be well aware of these restrictions.

Equivalent Isotropically Radiated Power

Equivalent isotropically radiated power (EIRP) is the output power of the antenna. EIRP is also regulated by the FCC, as it determines the amount of power distributed by the intentional radiator and the power gain of the antenna. For example, if a transmitter is connected to a 10-dBi antenna (which takes the RF signal and amplifies it 10 times) and this is added to the 100 mW delivered from the intentional radiator, the EIRP is 1000 mW, or 1 W.

The four areas of power calculations in WLAN are

1. Transmission power
2. Loss and gain between the radio and the antenna
3. Intentional radiator
4. Antenna power gain

Each of these calculations is important to understand to stay within the FCC restrictions. Incidentally, the wireless device manufacturer should be cognizant of this information, so when deciding on the right wireless vendor, communicate your concerns about these restrictions and whether or not the vendor has followed the applicable FCC rules and regulations.

Watts and Milliwatts

The basic unit of power is the watt (W) as defined as 1 ampere (A) of the current at 1 volt (V):

$$1W = 1A \times 1V.$$

A single lightbulb, plugged into a standard 120-V outlet, generates about 7 W of power. That single bulb can be seen for miles with a clear LOS. The FCC ruling for the unlicensed RF spectrum is 4 W of power to be radiated from an antenna. This may not seem like much, but RF isn't about illuminating a lightbulb. Its purpose is to send data and 4 W is enough to send a signal with a clear LOS for miles.

A milliwatt (mW) is 1/1000 of a watt. Most power levels for RF are defined in milliwatts and decibels.

Decibels

Decibels (dB) are an important component for RF measurement because a receiver antenna can pick up an RF signal as small as 0.000000001 W, which is a cumbersome number and means nothing to the average administrator. Decibels make those tiny numbers more manageable and understandable. Decibels are a logarithmic relationship to the linear power measurement in watts. In RF, a logarithm is the exponent to which the number 10 must be raised to reach the desired value.

For example, the logarithm (log) of 1000 is the number 3 because $10^3 = 1000$ ($10 \times 10 = 100$, $100 \times 10 = 1000$, $10 \times 10 \times 10 = 10^3$). On most linear scales of measurement the reference is fixed at zero: 0 W = no power, 0 miles per hour = no movement. With logarithms, however, you can't have a negative value or a value of zero, because decibels are a relative measurement, not absolute (like watts).

Power Gain and Loss

Power gain and loss are relative concepts so they're measured in decibels and not milliwatts. Although the gain and loss in RF may reference absolute power measurements, losing half the power in an RF system refers to losing 3 dB (again, relative, not absolute). When a system loses half its power, measured as −3 dB, and then loses another half of its power, which is another −3 dB, the total loss is 3/4 of the original power. The system first loses 1/2 of its power, then another 1/2, which is really considered 1/4 of the original power, so the total loss is 3/4 of the original power.

Here's a quick reference, called the 10s and 3s of RF mathematics:

−3 dB = half the power in mW
+3 dB = double the power in mW
−10 dB = 1/10 the power in mW
+10 dB = 10 times the power in mW

Power gain or loss can usually be measured by dividing the amount of gain and loss by 10 or 3, or even both.

dBm

This is the equation used to convert mW into dBm:

$$P\text{dBm} = 10\log P \text{ mW}.$$

dBm (decibels below 1 mW is used to measure RF power distribution. The "m" in dBm refers to the fact that 1 mW = 0 dBm; thus dBm is a measurement of absolute power and not relative dB power. Remember that when using dB, it's a relative measurement; so 10 mW + 3 dB = 20 mW because +3 dB is double the value (and −3 dB is half the value). This rule will help with quick calculations when defining power levels in dBm and dB measurements.

If we take the 10 and 3 rule to a power measurement of −26 dBm, we get −10 dB − 10 dB − 3 dB − 3 dB. Beginning at the reference of 1 mW, using the logarithm of 26 divided by 10 twice and divided in half twice (remember −3 dB means 1/2), we get

$$1 \text{ mW}/10 = 100 \, \mu\text{W (microwatts)},$$

$$100 \, \mu\text{W}/10 = 10 \, \mu\text{W},$$

$$10 \, \mu\text{W}/2 = 5 \, \mu\text{W},$$

$$5 \, \mu\text{W}/2 = 2.5 \, \mu\text{W},$$

such that −26 dBm equals 2.5 µW of power.

dBi

The decibel measurement for power gain of an antenna is dBi, whereas the "i" represents isotropic, referring to the change in power from the isotropic radiator in all directions. As with dB, dBi is a relative measurement. For example, connecting a 10-dBi antenna to 1 W of power equals 1 W + 10 dBi (or 10 times the power) = 10 W. Like dB, dBi is a relative unit of measurement that can be added or subtracted from other decibel units.

Figure 5-6 depicts a typical radio connected to an external directional antenna. There are two connectors and a cable that link the antenna to the radio. Table 5-2 illustrates the method for determining how to calculate the power loss from the connectors and cable and the power gain from the antenna. As an example, the 100-mW radio will lose −3 dB of power from the first connector (Point A), which means half of its power, down to 50 mW.

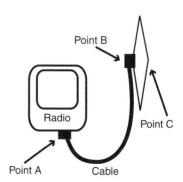

FIGURE 5-6 Outdoor radio connected to a flat panel antenna with a coaxial cable.

Table 5-2 Determining Power Output

Radio (mW)	Point A	Point B	Point C
100	−3 dB	−3 dB	+10 dBi
=100	Divided by 2		
=50		Divided by 2	
=25			×10
=250			

The 50 mW will also be cut in half based on the −3-dB loss from the second connector, down to 25 mW. The LMR600 cable is about 4 feet long, which has a signal loss of about 4 dB at 2.4 GHz, or 7 dB at 5.8 GHz *per 100 feet*, so 4 feet is minuscule. The antenna is a 10-dBi directional panel antenna that multiplies the remaining 25 mW of power by 10, making the EIRP 250 mW.

Antenna Radiation Patterns

An antenna radiation pattern is a three-dimensional radiated area in either an elevation pattern and/or the azimuth pattern (see Figure 5-7). The elevation pattern is a graph of the energy radiated from the antenna looking at it from the side and the azimuth pattern is a graph of the energy radiating from the antenna looking down from above the antenna.

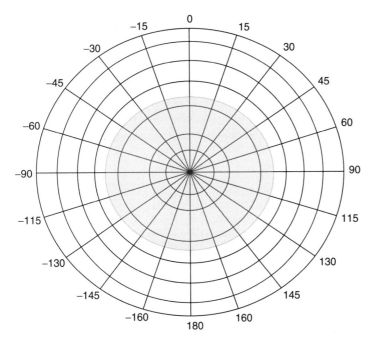

FIGURE 5-7 The azimuth pattern for a dipole omni-directional antenna.

Polarization

Antenna polarization is the orientation of the traveling electromagnetic waves to and from the source. The most common polarizations for a digital video security (DVS) application are linear, typically either vertical or horizontal, and circular, which includes circular right hand (RHCP), circular left hand (LHCP), elliptical right hand, and elliptical left hand. There are also antennas that provide a combination of these orientations, or even dual orientation.

In cellular and other wireless applications, signals are typically vertically polarized. This means that antennas must be vertically polarized for the optimal signal performance, or depending on the antenna gain, no signal at all. Weather and multipath interference can depolarize RF signals, reducing the power gain and deteriorating the signal.

Antenna Types

Although there are many different types of antennas, there are a select number of designs most relevant to 802.11x DVS applications. Antennas are also frequency-specific: a 700-MHz antenna won't work if the radio is set to 4.9 GHz. Antennas must match the frequencies they transmit and receive.

Antenna types include:

1. Dipole
2. Multiple element dipole
3. Yagi
4. Flat panel
5. Parabolic dish

Dipole Antenna

Commonly known as an omni-directional, all dipole antennas have a generalized radiation pattern. The donut-shaped elevation pattern shows that a dipole antenna is best used to transmit and receive from the broadside of the antenna, and is very sensitive to matching horizontal positioning and any movement away from a perfectly vertical position. At about 45° from perfect verticality, the omni's signals, both received and transmitted, will degrade to more than half.

Physically, dipole antennas are cylindrical and limited in power gain due to their widespread coverage and are used most often in mobility applications. The dipole antenna isn't a directive antenna, as its power is radiated 360° around the antenna (one of the reasons for FCC power gain limitations). Dipole antennas are also the most common culprit in interference issues because of their widespread radiated pattern. A mobility device requires a dipole antenna since there's no way to tell where the next AP will be for connectivity. If a mobile unit discovers an AP north of its current position, the antenna continues to radiate 360° in all directions, creating noise and/or interference for any other AP in the area attempting to use the same frequencies and channels.

Multiple Element Dipole Antennas

Multiple element dipole antennas have similar characteristics as the dipole, but with added directionality in the elevation pattern and additional power gain as a result of using multiple elements.

Multiple elements can be configured with different power gain, allowing for multiple antenna designs with similar physical characteristics. Instead of radiating equally in all directions on the horizontal plane, like the dipole antenna, the multiple element dipole antenna can focus its additional power using directivity.

Yagi Antennas

Yagi antennas (see Figure 5-8) include an array of independent antenna elements with only one of the elements driven to transmit electromagnetic waves. The number of elements (specifically, the number of director elements) determines the gain and directivity. Yagi antennas aren't as directional as parabolic dish antennas, but are more directional than flat panel antennas.

Flat Panel Antennas

Flat panel antennas are shaped like a square, diamond, or rectangle and are laser-focused in both the vertical and horizontal planes (they can also be vertical and horizontally polarized). Figure 5-9 depicts the azimuth radiated pattern for a 21-dBi flat panel antenna with a vertical radiated plane of 10° and a horizontal plane of only 8°, making it an optimum choice for long distances and its ability to avoid interference and noise in heavily wireless locations.

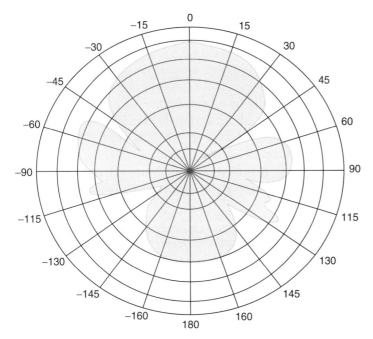

FIGURE 5-8 Yagi azimuth radiated pattern.

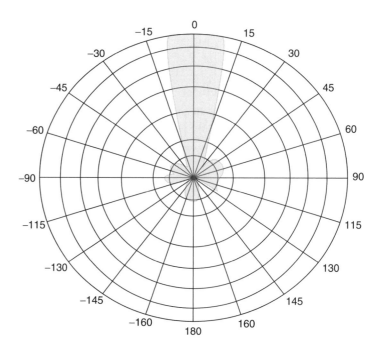

FIGURE 5-9 Flat panel antenna azimuth radiated pattern.

Parabolic Dish Antennas

Parabolic dish antennas use special design features and multiple elements to achieve very high gain directivity. These antennas use a reflective dish shaped like a parabola to focus all received electromagnetic waves on the antenna to a single point. The parabolic dish also works to catch all the radiated energy from the antenna and focus it in a narrow beam for transmission with very high power gain.

Sector Antennas

Sector antennas are one of the most common antennas, as they provide a power gain close to directional antennas but with a wider beam width. Directional antennas radiate power in a particular direction that's usually a ratio and/or angle of more laser-focused radiation intensity. Although a sector antenna may be considered a directional antenna, it can focus its signal within 30°, 45°, 60°, 90° or 120°, in one direction horizontally (see Figure 5-10) and it has a wider spread vertically than a flat panel or parabolic antenna (see Table 5-3).

WLAN Standards

802.11

The initial version of the 802.11 wireless standard (1997) used FHSS or DSSS and differential binary phase shift keying (DBPSK) modulation. The spread spectrum radio transmission was a physical layer development to prevent radio signals from being monitored

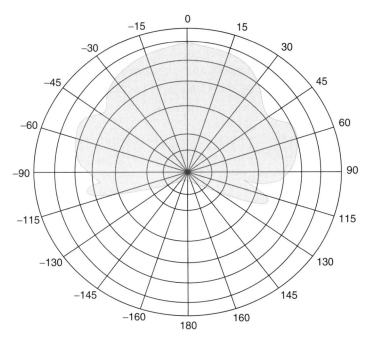

FIGURE 5-10 Radiated azimuth pattern for a 90° sector antenna.

Table 5-3   Antenna Comparison

Antenna	Radiation Patterns (Degrees)	Power Gain	Polarization	Note
Dipole	360	Low	Linear	Omni
Multiple element dipole	360 + second element	Low/medium	Linear	Omni and Directional
Flat panel antenna	10-30	Medium	Linear/circular	Directional
Parabolic dish antenna	3-10	High	Linear/circular	Directional Long Range
Yagi antenna	Endfire	Medium/high	Linear	Omni Long Range
Sector antenna				
Slotted antenna	Broadside	Low/medium	Linear	
Microstrip antenna	Endfire	Medium	Linear	

or blocked. It was developed by creating a transmitter and receiver that change frequencies quickly and regularly to stay a step away from any uninvited hackers. The DSSS method of transmission uses a subsignal through the entire band rather than hopping from one frequency to another. DSSS offers higher data rates and shorter delays than FHSS, because there's no re-tuning involved.

Both FHSS and DSSS are resistant to interference from conventional radio transmitters. Because the signal doesn't stay in one place on the band, FHSS can elude a jammer (a transmitter designed to block radio transmissions on a given frequency). DSSS avoids interference by configuring the spreading function in the receiver to concentrate the desired signal, but simultaneously spreads out and dilutes any interfering signal. DSSS avoids interference from fixed narrowband frequencies, but does have a problem if there are radios using FHSS as both will eventually hop onto the same frequency.

It's important to mention that the industrial, scientific, and medical (ISM) bands – which include the 2.4-GHz Wi-Fi, Bluetooth, telephone, and microwave ovens – and the 5-GHz Unlicensed National Information Infrastructure (UNII) bands don't require any license. Anyone with a wireless device can go wireless, anywhere at anytime. Regarding the 802.11 frequencies, the FCC Technical Advisory Committee (TAC) stated, "The TAC will be monitoring the consequences of an unplanned real-time experiment of uncoordinated spectral sharing with incompatible etiquette rules." In other words, "Good luck."

The wireless MAC layer protocol includes the carrier sense multiple access/collision avoidance (CSMA/CA). This is the channel access mechanism used by WLANs in the ISM bands that listens before transmitting data and will back off if a collision is detected. The basic principles of CSMA/CA are an asynchronous message (con-nectionless) delivering a best effort service with no bandwidth, no latency guarantee, and suited for TCP/IP networks. CSMA/CA doesn't come from the wireless world, but has its roots in CSMA/CD instead, which is the foundation of Ethernet. The primary difference between the collision detection of wired LANs and collision avoidance of wireless LANs is that the transceiver can listen while transmitting on a wire, but can only avoid collisions because its own transmission strength would mask all other signals.

CSMA/CA initiates a back-off procedure when detecting interference. This can be det-rimental to streaming video (depending on the fps and bitrates) as it reduces the power and thus the bandwidth of the signal, causing latency and stuttering video images. This is a major issue with using wireless radios that follow the Wi-Fi standard, designed for wireless connectivity for sharing e-mail, intranet resources, and surfing the Internet, but that can all be done asynchronously without any major repercussions. DVS is about real-time streaming video, which must be synchronized with the VMS server and archiver. This is where the specifications of the radios become more important.

802.11b

Higher DSSS rates of 5.5 and 11 Mbps were introduced in 1999. This popular 802.11b (2.4-GHz Wi-Fi band) operates between 2400 and 2483.5 MHz, requiring a frequency bandwidth of approximately 22 MHz for each channel. The channel boundaries are defined as ±11 MHz from center frequency, with the signal 30 dB lower than at center. To provide these higher data rates, 802.11b uses a complementary code keying (CCK) modulation technique that makes better use of the frequency.

There are only three non-overlapping channels in the 2.4-GHz spectrum (see Figure 5-11). This means that anyone using Channels 1, 7, and 13 would be safe from

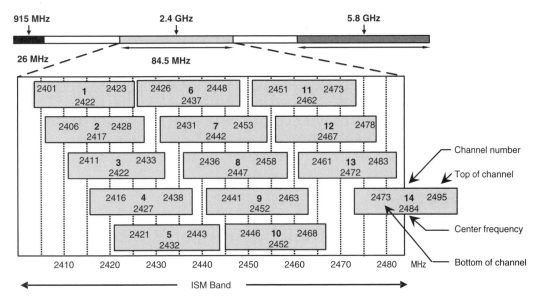

FIGURE 5-11 2.4-GHz channels.

interference from each other, but using Channels 1, 2, and 3 will cause interference because they overlap within the band.

802.11a

Also introduced in 1999 was the new high-bandwidth 802.11a standard, which wasn't compatible with 802.11b. The 802.11a is a physical layer (PHY) standard that operates in the 5-GHz UNII band using orthogonal frequency division multiplexing (OFDM), with data rates of 6-54 Mbps (see Table 5-4). The OFDM is composed of a 20-MHz channel bandwidth encapsulating 52 independent subcarriers based on data rate and error correction techniques. The higher the bandwidth, the more complex the error correction and more fragile the signal becomes.

Since the 802.11a standard isn't backward compatible to 802.11b, it's rarely used in the field. This is beneficial when deploying video surveillance, as 802.11a provides bandwidth up to 54 Mbps and rare usage, which would limit interference and channel-hopping by other radios. However, the 5-GHz spectrum is limited by FCC regulations to only half the power of a 2.4-GHz band and is more highly attenuated by distance and walls.

OFDM

The OFDM spread spectrum technique takes a signal and transmits multiple signals simultaneously, rather than one signal transmission (see Table 5-5). OFDM breaks down the frequency into smaller 1-MHz frequencies and each signal travels within its own unique frequency range or carrier.

Table 5-4   The 5-GHz Band for the United States, Europe, and Asia

Frequency	Channel Number	FCC (GHz)	ETSI (GHz)	ASIA (GHz)
Lower band (36 default channel)	34	-	-	-
	36	5.180	5.180	-
	38	-	-	-
	40	5.200	5.200	-
	42	-	-	-
	44	5.220	5.220	-
	46	-	-	-
	48	5.240	5.240	-
Middle band (52 default channel)	52	5.260	-	-
	56	5.280	-	-
	58	5.300	-	-
	60	5.320	-	-
H band	100	-	5.500	-
	104	-	5.520	-
	108	-	5.540	-
	112	-	5.560	-
	116	-	5.580	-
	120	-	5.600	-
	124	-	5.620	-
	128	-	5.640	-
	132	-	5.660	-
	136	-	5.680	-
	140	-	5.700	-
Upper band (149 default channel)	149	5.745	-	5.745
	153	5.675	-	5.675
	157	5.786	-	5.786
	161	5.805	-	5.805
ISM band	165	5.825	-	-
			-	-
			-	-
			-	-
			-	-
			-	-

Table 5-5   WLAN Standards

802.11 Protocol	Release	Frequency (GHz)	Mbps	Modulation
	Jun 1997	2.4	002	DSSS
a	Sep 1999	5	054	OFDM
b	Sep 1999	2.4	011	DSSS
g	Jun 2003	2.4	054	OFDM
n	~Nov 2009	2.45	600	OFDM

The spacing between the frequencies prevents the demodulators from bleeding over frequencies for higher spectral efficiency, resiliency to RF interference, and lower multipath distortion. RF terrestrial broadcasting includes multipath interference when the transmitted signals are received from various paths, at different lengths, and at different times. When these multiple versions of the signal interfere with each other (inter-symbol interference; ISI) it becomes difficult to extract important data. OFDM is sometimes called multicarrier or discrete multitone modulation and is used as the digital modulation technique for digital TV in Europe, Japan, and Australia. It's quite effective at streaming digital video across RF.

The IEEE 802.11a standard outlines the use of OFDM in the 5.8-GHz band.

802.11g

The 2003 standard 802.11g is a combination of 802.11a and 802.11b dedicated to use of OFDM instead of DSSS as a basis for achieving the higher data rates of 6, 9, 12, 18, 24, 36, 48, and 54 Mbps. Although 802.11g has the same data rates and modulation types as 802.11a, it uses the 2.4-GHz frequency band and even shares the same channels as 802.11b. This provided backward compatibility with 802.11b devices with duplicate speeds and modulation types, and by using quadrature phase shift keying (QPSK) for 2 Mbps and binary phase shift keying (BPSK) for 1 Mbps.

Although you can mix 802.11b and 802.11g clients with a single 802.11b or 802.11g AP, there may be a performance problem depending on the demand of bandwidth. While 802.11g defers to its legacy brother, 802.11b doesn't return the favor. This can cause collisions degrading all traffic, while the 802.11g mode protects traffic by using a CTS-to-self function. This function sends a broadcast CTS message at an 802.11b device to transmit a packet and clear the channel.

802.11n

The 802.11n standard is a high-bandwidth solution that uses multipath interference to its advantage. The radio can now "read" multipath interference at the receiver end and determine that these microwave signals are actually part of the data delivered by the transmitter. The radio is now smart enough to put those scattered pieces together and offer bandwidth of up to 300 Mbps by using dual radios and special multiple antennas.

It's becoming clearer that 802.11a/b/g WLAN can't provide adequate performance and/or bandwidth for many of today's networking applications. In response to this need, both IEEE Task Group "N" and the Wi-Fi Alliance have set new expectations for the next WLAN generation. In July 2003, the 802.11n task group was formed to create a new standard that provides a throughput of at least 100 Mbps. A number of proposals were made that included the use of MIMO-OFDM, 20- and 40-MHz channels, and packet aggregation techniques. These were the stepping stones for the first 802.11n standard draft in 2005.

The 802.11n standard defines a range of mandatory and optional data rates in both 20- and 40-MHz channels. Other optional features include reducing the guard intervals in half to increase the maximum data rate for dual spatial streams in a 40-MHz channel up to 300 Mbps and three or four spatial streams to reach 600 Mbps.

MIMO

MIMO, also referred to as smart antenna systems, corresponds to the number of antennas used to carry data in 802.11n. MIMO divides a data stream into multiple unique streams, each modulated and transmitted through a unique radio-antenna chain, concurrently using the same frequency channel. This revolutionary technique leverages environmental structures and takes advantage of multipath signal reflections to actually improve radio transmission performance instead of being detrimental "multipath interference." The hefty bandwidths provided by 802.11n are achieved by using multiple antenna systems for both transmitter and receiver, and each antenna provides both functions.

MIMO provides two very important benefits: antenna diversity and spatial multiplexing. By using multiple smart antennas, MIMO resolves information from multiple signals using spatially separated receive antennas. MIMO also spatially resolves multipath signals, providing diversity gain, thus improving a receiver's ability to recover intelligent data (see Figure 5-12).

MIMO requires a separate RF chain and analog-to-digital converter (ADC) for each MIMO antenna. Implementation of 802.11n and MIMO can be as much as six times the cost of installing a single radio and antenna because the radios are more costly, the antennas are proprietary to the system and also more costly, and six antennas are needed instead of two. Before you implement this expensive plan, be certain that your requirements call for more than 54 Mbps.

The other advantage of 802.11n and MIMO is the intelligence in data recovery and how it uses multipath. This is perfect for the RF non-line-of-sight (nLOS) scenarios. There's no advantage other than bandwidth to clear LOS conditions, but if the implementation requires RF transmissions around corners of buildings, through a forest of trees, or other such nLOS situations, the multipath advantage gained by MIMO technology will be well worth the cost of implementation.

The advanced digital signal processing (DSP) hardware makes it possible to decipher the multipath differentiated RF signals even though they're all on the same frequency, scattered, untimely, and from different directions.

MIMO Antennas

As mentioned before, MIMO antennas are unique both in design and implementation. If there are three transmit antennas and three receive antennas, the configuration is often referred to as 3 × 3 MIMO or 2 × 3 MIMO. The first number is the number of transmit antennas and the second number is the number of receive antennas.

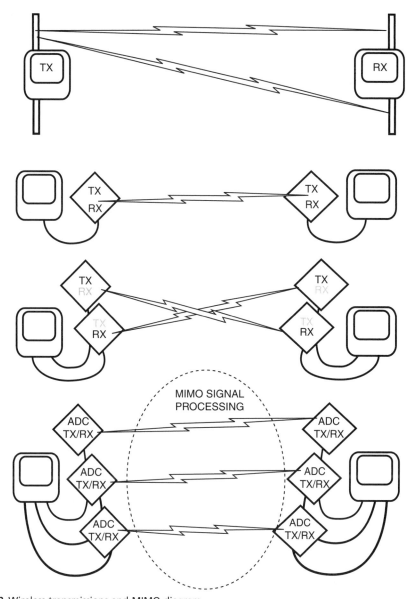

FIGURE 5-12 Wireless transmissions and MIMO diagram.

802.11s – Mesh

The 802.11s wireless mesh networking standard is based on the 802.11 WLAN standard, and has been actively explored for a few years. Meanwhile, mesh networking is proliferating in many different applications, including DVS. The IEEE 802.11s task group is attempting to craft an industry-wide standard that creates cohesiveness among vendors, as currently each vendor has its own "special sauce." Without vendor cohesive-ness, new protocols are introduced that are proprietary to each individual manufacturer, thus preventing any interoperability. Several 802.11s standard drafts have been released, but many issues remain unresolved. A detailed study on the existing 802.11s standard is available at www.ieee.org.

Although wireless mesh networking is in its infancy relative to wireless networking in general, it's a functionality that makes wireless networking even more attractive and something that can't be done within a wired network.

Wireless Mesh Networking

A wireless mesh network is made up of two or more wireless radios working together to share routing protocols to create an interconnected RF pathway. A wireless mesh net-work, no matter how many radios are included, creates only a single name identifier (SSID) and could also create a single IP address for the entire mesh. This clearly distinguishes the mesh from another wireless or mesh network.

Wireless mesh networking includes three types of topologies, based on requirements and LOS:

1. Point-to-point
2. Point-to-multipoint or multipoint-to-point
3. Multipoint-to-multipoint

While point-to-point (PtP) and point-to-multipoint (PtMP) network topologies have been the standard for fixed wireless deployments, mesh networking has overcome some disadvantages in the traditional wireless topologies.

Point-to-Point

Mesh radios can provide a simple PtP solution (similar to a laptop radio connecting to an AP for shared Internet access) by creating a wireless backhaul between two wired Ethernet locations, or adding a security camera from a remote location with only power available, into the DVS system. PtP solutions offer an option for any location regardless of data connectivity.

A PtP network is simply a wireless network where two radios, each with high gain antennas, directly communicate with each other to provide high-performance, high-bandwidth dedicated connections. These links are quick to deploy individually, but don't easily scale to creating a large network.

Point-to-Multipoint

There's also a PtMP network, which includes a number of wireless radios pointing to a centralized uplink radio with access to a fiber or Ethernet network. This type of network is easier to deploy than PtP, simply by adding a new radio node to the existing "spoke and wheel" topology. The new radio node obviously must be within the signal range, with clear LOS of the uplink base station. Trees, buildings, poles, and other LOS obstructions make PtMP difficult within urban environments. A PtMP network provides an uplink backhaul, high-speed connection to the wired network and onto the VMS software and archives.

Before delving deeper into mesh networking, it's important to make clear that not all mesh networking equipment is alike. The lack of a current standard has allowed a plethora of individual, proprietary solutions to flourish, and the open experimentation has created some failures as well as some outstanding products that add to the magic of wireless.

Mesh

The primary benefit of a wireless mesh network is the extended bandwidth and redundancy. A wireless mesh radio can only communicate with another wireless mesh radio using similar protocols such as SSID, end-to-end encryption, wireless encryption, etc. But mesh radios, although they may use Wi-Fi chipsets, don't function as wireless AP and only communicate with other wireless mesh radios in that SSID and mesh ID network. It provides added security; even more than a wireless camera or AP.

This works because mesh radios, which provide the same bandwidth as 802.11a/g radios, become an RF extension of a wired network by providing a link in and out of an Ethernet network. The mesh radios are invisible to the end user – just another Layer 2 switch without wires. As seen in Figure 5-13, there's redundancy when the signal between two radios is interrupted, corrupted, or the radio loses power and the mesh network "self-heals" by redirecting the traffic through another radio in the "mesh."

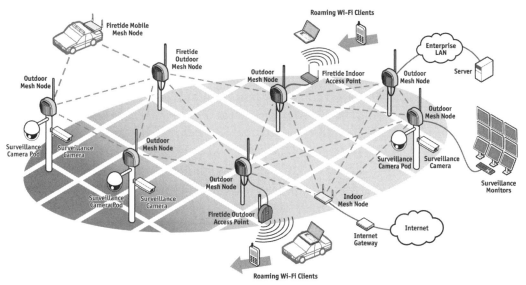

FIGURE 5-13 Besides redundancy through multiple paths and self-healing capabilities of wireless mesh networking, the very nature of mesh networking provides mobility capabilities that allow vehicles to join each mesh when necessary.

Wireless Security Options and Considerations

There's no such thing as 100% digital security, wireless or otherwise. Computers are getting faster exponentially with more memory and cheaper hard drive space. Breaking a 128-bit key is just a matter of time. However, it's not the 128-bit key encryption that threatens the entire digital world; it's the layers of security before it and behind it.

Working in downtown Chicago, if I run my Netstumbler (excellent wireless tool and a free download at www.netstumbler.com/downloads), I could, depending on the location, pick up almost 100 APs and/or wireless routers for individual homeowners, offices, and institutions providing wireless connectivity. Many of those wireless APs don't have any encryption enabled and some are still using the default ID and password. Even more baffling, these APs are set up using dynamic host configuration protocol (DHCP), so anyone who attempts to connect to the AP and the internal network will be provided a free, dynamic IP address and a welcomed handshake into the network.

A wireless AP works great when there's a desperate need for Internet access, but leaves the homeowner somewhat baffled at the crawling rate of that access. It's easy enough to look around and see that someone somewhere has an AP wide open to the world. Getting the default ID and password is as simple as identifying the MAC addresses if you're unable to see the brand name in the SSID (e.g., Linksys). Each manufacturer has a unique number recognizable in the first digits of the MAC address. This is where digital security breaks down. If you're going to use a computer, you need to know the rules.

Here are my rules for digital security:

1. Don't save passwords and IDs in the Web browser. It's best to write them on a piece of paper next to the computer because you can control who comes into your home or office. There's a reason there are new security service packs and patches released for Web browsers.
2. Implement a firewall or at least a router in front of the Internet (don't use the default ID and password). A $60 router, wireless or otherwise, automatically closes the 65,000 incoming TCP and UDP ports that are wide open when a computer is plugged directly into a cable or DSL modem.
3. Turn off the computer when not using it.

Here are my rules for wireless security:

1. Change the default ID and password in the AP and/or wireless router for the same reason I discussed above.
2. Turn on the maximum WPA/WEP encryption available and keep it on at all times (see Table 5-6).
3. If the AP and/or router offers the ability to save the configuration (and encryption key with the configuration file) use a random alphanumeric encryption key (not your

Table 5-6 Wireless Security Encryption

Standard	Data Rate	Modulation Scheme	Security
IEEE 802.11	Up to 2 Mbps in the 2.4-GHz band	FHSS or DSSS	WEP and WPA
IEEE 802.11a (Wi-Fi)	Up to 54 Mbps in the 5-GHz band	OFDM	WEP and WPA
IEEE 802.11b (Wi-Fi)	Up to 11 Mbps in the 2.4-GHz band	DSSS with CCK	WEP and WPA
IEEE 802.11g (Wi-Fi)	Up to 54 Mbps in the 2.4-GHz band	OFDM above 20 Mbps, DSSS with CCK below 20 Mbps	WEP and WPA
IEEE 802.16 (WiMAX)	Specifies WiMAX in the 10-66-GHz range	OFDM	DES3 and AES
IEEE 802.16a (WiMAX)	Added support for the 2-11-GHz range	OFDM	DES3 and AES
Bluetooth	Up to 2 Mbps in the 2.45-GHz band	FHSS	PPTP, SSL, or VPN

dog's name) and save the file in a safe place. A random number provides better protection than something someone may know about you.

4. Change the encryption key every few months or so.
5. Change the default SSID (so it's more difficult to determine the hardware used) or if applicable, disable the SSID broadcast, which will make the network invisible (make sure accessibility is set up on the devices with access).
6. Enable MAC address filtering. APs and wireless routers provide a means of allowing only select devices to gain access to the wireless network. This is done by adding the physical MAC address of each computer with acceptable access.
7. Disable DHCP and use only static IP addresses.

Wireless Everything

The "wireless everything" security camera (using solar power) was a Lorex SG8840 wireless IP camera (with a built-in lithium battery for when the sun goes down), connected to a receiver sitting indoors, linked to a Verint digital video encoder, and viewed through the VMS software at CIF and 4CIF resolutions. The bandwidth capabilities were actually quite impressive.

Wireless IP cameras have become another client on the WLAN and like a laptop can be configured with encryption, authentication, and accessibility from anywhere there's connectivity to an access point. My first "totally wireless" experiment included a Linksys/Cisco WVC54GCA wireless IP camera (using solar power), which linked to my Linksys wireless AP. This was then linked wirelessly to my laptop. However, the Linksys/Cisco WVC54GCA didn't have a backup lithium battery, so when the sun went down, so did the camera. More important, the camera was incompatible with the VMS software, so I was locked into using the camera's Web interface.

Channel Planning

The RF spectrum for non-licensed frequencies, used by all WLAN implementations, is relatively small, depending on the number of wireless networks actively used within the area. If a single DVS implementation is the only wireless user of the 5.8-GHz spectrum, then there's no need to consider a channel plan with the neighbors, but if there are dozens of other companies, institutions, or even homeowners using the same spectrum, then finding an open channel to use, without any interference, can be a problem.

For example, Office A has a wireless IP security camera currently linked to a Linksys network and has an AP using channel 1. A neighbor (Office B) decides that his other neighbor's (Office C) wireless channel is bleeding too much into Office B's channel. So, Office B (or the radio, on auto-channel select) decides to choose channel 1 because at this very moment Office A's wireless signal faded because someone in Office A is using the microwave oven in the cafeteria. Now you have Office B's AP (which incidentally is a newer model, and it happens to be raining) playing dueling frequencies with Office A and Office C, and that knocks the camera off the WLAN. This is why a channel plan is important. The IT directors from Offices A, B, and C can have a meeting and decide to each choose a specific channel and use only that channel with the hope that no new users decide to use the same frequency in the area.

Wireless takes the digital data and converts them back into an analog RF signal, which is then received and converted back into digital data for the LAN. The wireless radios take the digital video signal and transmit it to a receiving radio, which is then plugged back into the network and into the VMS software. Wireless network video gives you the flexibility to extend your video network beyond wires, walls, and buildings.

Configuring Access Point Radios

Although each manufacturer's radio design is different, the basics of configuration for an AP are the same. If the following steps seem vague, it's because they're a simple example of AP configuration. Nevertheless, all WLAN radios, whether they're AP, wireless routers, or mesh radios, typically have two LED status lights to let you know that there's power, and the other to notify the user that the radio(s) is active (usually as a STATUS light).

Assuming that the AP includes a Web interface, providing each with access through Ethernet, the steps for accessing the device are similar to a digital video encoder or IP camera. The goal is to gain access to the built-in Web server that hosts the device management pages.

Set the network adapter of the configuration laptop to the same subnet as the default IP address of the AP and then launch the Web browser. Next, type http://defaultipaddress into the address location of the Web browser. Then you'll be prompted to log in using the default ID and password.

If you're prompted to follow a Configuration Wizard (see Figure 5-14), click NEXT and follow the instructions, or click on EXIT to go to System Status. Somewhere on the page

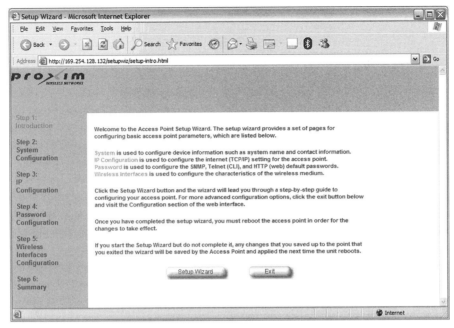

FIGURE 5-14 Configuration Wizard.

you should find a way to navigate changing the ID and/or password. Once that's accomplished you may need to reboot. Upon returning, move to the Configure and/or Network navigation button. Follow the navigation until you're presented with the option of entering an IP address. Disable the Access Dynamic Address or DHCP and choose Static Address or navigate to Assign an IP Address. Enter the intended IP address of the access point and choose Radio settings. Under the Radio settings, choose the desired frequency to use a fixed channel (or Auto Channel Select if available). Enter a new SSID for the AP and then choose the Security tab or navigation button. Choose Enable Security and select the highest encryption rate available. Once prompted, enter a secret key code and choose OK and/or SAVE. You may need to reboot the AP for it to completely write the changes into flash memory.

Configuring a Mesh Radio

Firetide mesh technology deploys an encapsulated configuration of a wireless mesh. The number of radios you wish to use within a specific mesh determines how best to configure the overall mesh. For example, if you're using five radios within a mesh, simply power up all five radios with the supplied antennas and they will immediately communicate with each other to form a default mesh (see Figure 5-15).

The POWER status LED will immediately turn green, indicating that the radio is going through its boot cycle. After a few moments, the STATUS light turns green, indicating

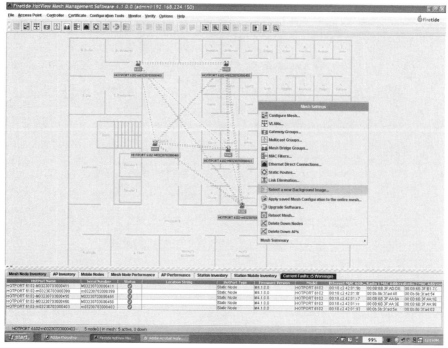

FIGURE 5-15 Firetide mesh initial configuration screen.

that the radios are operational. Once the RADIO 1 and/or RADIO 2 lights turn green (Firetide mesh radios come with single or dual radios) the radio is communicating in its default mesh.

Any radio that's unable to join (straight out of the box) may be defective, so you may need to attempt a manual recycle of power to see if it boots up and joins the mesh network. This may take a few minutes.

Network Management Software

Firetide meshes are IP-independent and will transport Ethernet packets using any IP addressing system within the enterprise. For management purposes, Firetide meshes are assigned IP addresses that exist only for maintainability through the network management software.

A Firetide mesh is a self-running entity; it doesn't require a network management system (NMS) for moment-to-moment operation, but the configuration process does require the proprietary Firetide HotView Pro software to initiate functionality. Firetide mesh radios aren't APs (although Firetide does manufacture wireless AP units) where the AP becomes a bridge between the wireless radio within a laptop and the LAN. Firetide mesh radios only communicate with other Firetide radios and the only

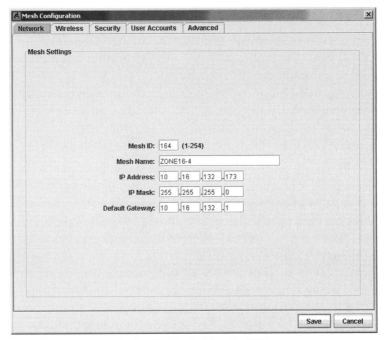

FIGURE 5-16 Initial configuration screen in the Firetide HotView Pro NMS.

method of configuration to access the mesh is by using HotView Pro NMS. Accessibility to that mesh, once imported into the HotView Pro NMS, is limited to that individual NMS.

Once HotView Pro is installed, launch the NMS and the default mesh will appear (see Figure 5-16). If the default mesh doesn't automatically appear, add it by navigating from FILE > ADD MESH and then type in 192.168.224.150 and the default ID and password. Right-click on the background and choose Configure Mesh from the drop-down menu, or press the Configure Mesh icon from the top menu to configure global mesh settings. Once presented with the Mesh Configuration dialog box, click on the Network tab. Next, type in a Mesh Identifier (Mesh ID), choosing a number from 1 to 255 – a unique number that can't be used on another mesh in the network. Type the Mesh Name, which warrants a naming convention using an alphanumeric representation of the location. This will make it easily recognizable later within the NMS.

Type in a unique IP address for the entire mesh; no matter how many radios are part of the designed mesh, it only uses a single IP address, which is somewhat unusual for wireless radio configurations. Each Firetide radio does have a unique MAC address and one for the built-in Ethernet switch, but the unique IP address is invisible. The IP Mask is 255.255.255.0 and the DEFAULT GATEWAY is the appropriate gateway assigned to the WLAN.

ESSID Encryption and Radio Settings

Click on the Wireless tab on the menu and type in a unique alphanumeric Extended Service Set ID (ESSID) field. Check the Enable ESSID Encryption box and choose the radio frequencies used to communicate with each radio.

Default is set to Bonded Mode, which links both radios together to double the bandwidth. Auto Channel Select is also an option, providing up to six different channels (frequencies) to use when hopping frequencies to optimize signal quality.

Security Settings

Choose the Security tab from the top menu and choose either a 128- or 256-bit encryption. There's a dual layer of encryption on the Firetide radios. The first is the encryption of the signal, which is then encapsulated with an end-to-end encryption (radio to radio). There's no way to join the mesh without these two encryption keys, and another Firetide mesh radio is required (Figure 5-17).

Choose User Accounts to change the ID and passwords used to access the mesh through HotView Pro, and Advanced settings for optimal configurations.

Click on SAVE at the bottom right and the units will automatically reboot while writing to flash memory. Once the configuration is completed change the IP address of the network adapter to match the new subnet settings assigned to the mesh to log-in again.

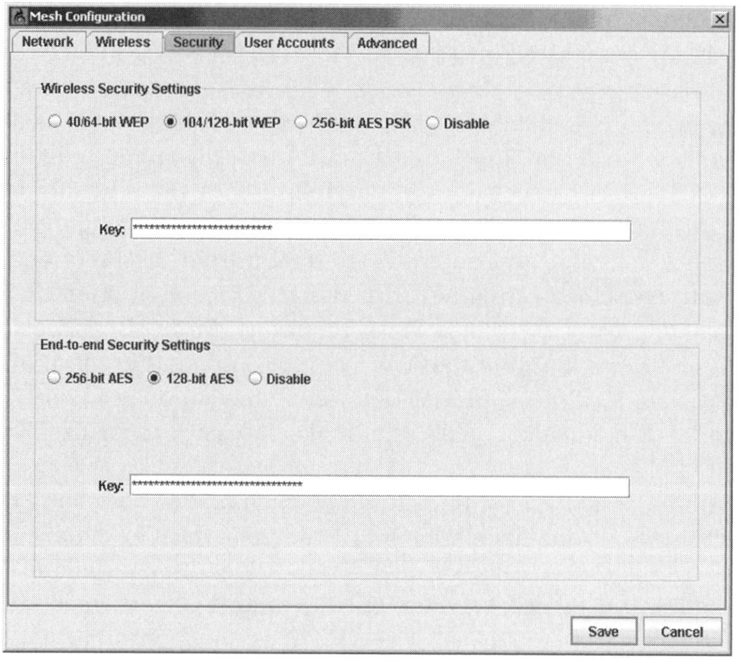

FIGURE 5-17 Firetide wireless and end-to-end security settings.

Wireless Antenna Coaxial Connectors

There are a few choices of coaxial connectors when using RF radios and antennas for a WLAN, and those choices are based on the coaxial cable chosen for the project. Unlike the coaxial cable choices for analog video, WLAN includes the choice of a higher quality cabling with lower impedance and loss. Figure 5-18 shows the optimal connector choices for antenna cabling. The first SMA connector is one of the more popular choices because it fits the more ordinary RG6, RG8, RG11, and LMR100 and 195 cabling (see the next section).

The N-type connector is used on the thicker cabling from LMR400 to LMR600. It's also available for the smaller sized cables. The recognizable BNC connector can also be used for antenna cabling, but an adapter is required as most wireless radios require the SMA, N-type, TNC, or the MMC and MMCX (which is typically used as an interim dongle from the radio to an SMA or N-type connector).

Antenna Coaxial Cables

Signal loss or attenuation happens all along the cable and connectors. The rule is always the shorter the cable, the better. Many consumer-use Yagis come with 18-22 feet of RG58 coaxial with 96% braid shielding, which includes about a 3.5-dB loss in signal (16-dB loss/100 feet/900 MHz). If your installation requires a longer cable run (more than 40 feet), upgrade to LMR600 (3-dB loss/100 feet). LMR600 may be thick and bulky, but it's the most durable for outdoor use and provides less loss for longer runs.

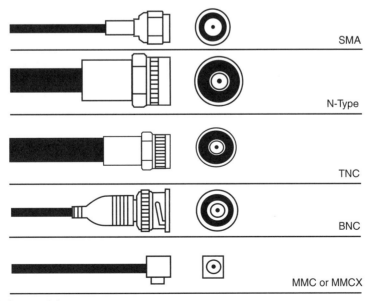

SMA

N-Type

TNC

BNC

MMC or MMCX

FIGURE 5-18 Wireless coaxial connectors.

The antenna needs to be mounted as high and in the clear as possible. However, there's a point at which increasing height to gain 1 dB can cause an extra 3 dB in cable loss. So careful consideration must be given when deciding if more height is needed and how much longer the cable will be. On the other hand, if excess cable length is present, remove it or run a long path to the antenna; never coil or fold up the cable.

There's a wide variety of coaxial cabling used for external antennas for wireless radios. The more popular brands include Times Microwave, Belden, Davis, and Andrew. The key to choosing the right cable for the project is based on requirements. Most of the projects I've worked on used LMR195, LMR400, and LMR600; each provides different specifications. The LMR600 is about 1/2″ thick as compared to the LMR195 (or RG58), which is 1/8″ thick. I recommend using LMR600 for an outdoor implementation because it's far more durable and because of its level of performance. While the LMR195 has a cable loss of 16 dB per 100 feet, the LMR600 has a cable loss of only 3 dB per 100 feet.

Remembering your calculations from earlier in this chapter (or you can go to www.timesmicrowave.com for a cable loss calculator), the LMR600 provides three times the efficiency of the LMR195.

Wireless Coaxial Termination

The key to properly terminating coaxial cable for use with wireless radios and antennas is having the right tools. There are two types of connectors, the screw-on and the crimp-on, much like the coaxial for video cabling. A crimp-on kit can cost anywhere from $100 to $500 depending on the type of connector (Figure 5-19).

LMR195/RG58 Cable Termination
Figures 5-20 to 5-24 illustrate guidelines for LMR195/RG58 cable termination.

LMR600 Termination
The crimp-on connector on the right in Figure 5-25 comes in three pieces: the connector, the crimp-on sleeve, and a heated rubber weatherized sleeve. If you don't have the heater to melt the rubber sleeve over the newly applied connector, use rubber and electrical tape. The EZ Connector on the left in Figure 5-25 can be applied with two crescent wrenches.

FIGURE 5-19 The required crimping tools. Left: for LMR195/RG58 and the SMA connectors; right: for LMR600 and the N-type connectors.

FIGURE 5-21 LMR195, RG58 cable SMA connectors are small and a challenge to handle.

FIGURE 5-20 Use an LMR195/RG58 stripping tool to begin termination (this tool also works well with Ethernet cables).

FIGURE 5-22 The SMA crimp-on connector slides underneath the braided insulation.

FIGURE 5-23 Before adding the SMA connector, make sure you have the metal sleeve in place for crimping.

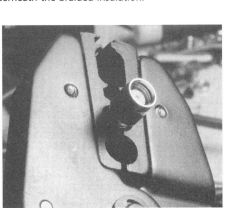

FIGURE 5-24 Have the right size crimping tool or you may damage the connector and/or cable. Once completed, weatherize the end with rubber or electrical tape.

FIGURE 5-25 LMR600 N-Type connector on the right is a crimp-on and the one on the left is an EZ Connector.

TERMINATING THE CRIMP-ON CONNECTOR

The two different types of LMR600 connectors each require a separate tool, but one stripper can work for terminating both ends. The goal when terminating these cables is to provide a perfect fit for the internal copper cable inside the connectors (see Figure 5-26), and I've found that filing the tip flat helps make the cable more effective.

The LMR600 cable stripper has two holes on each side. One side is about half as deep as the other and the shallower side is used first (Figure 5-27). Using the smaller side of the LMR600 cable stripper, remove the outer PVC sleeve and internal dielectric insulation to expose only the internal copper wire (Figure 5-28).

The opposite side of the cable stripper adds another 3/4″ to the outer PVC insulation sleeve to slide the crimp-on connector down to insert the internal copper cable into the hole inside the connectors (see Figure 5-28).

The copper inside the coaxial cable is inserted within the end connectors. Both connectors include the same hole for insertion (Figure 5-26).

Use a file to smooth down and flatten the internal copper cable for a snug fit (Figure 5-31).

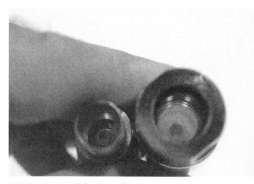

FIGURE 5-26 Inside the cable connectors.

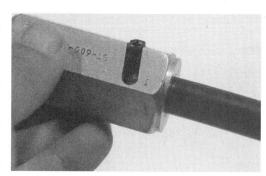

FIGURE 5-27 LMR600 Step 1.

FIGURE 5-28 LMR600 Step 2.

FIGURE 5-29 LMR600 Step 3.

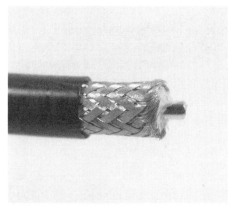

FIGURE 5-30 LMR600 Step 4.

FIGURE 5-31 LMR600 Step 5.

FIGURE 5-32 LMR600 Step 6.

FIGURE 5-33 LMR600 Step 7.

Push on the crimp connector until it's firmly in place (Figures 5-32 and 5-33).

Using the appropriate crimping tool, compress the metal sleeve so the connector is now permanently set to the cable end. Confirm that the end is firmly in place (Figure 5-34).

Confirm the solidity of the connector and then wrap the end with rubber or electrical tape.

Nothing brings down a camera and its digital video encoder like an antenna cable that isn't weatherized or a bent cable end. Bending the cable end (for LMR600) more than 90° pulls the internal copper cable out of the connector insert.

LMR600 TERMINATION EZ CONNECTOR

A major benefit of the EZ Connector is its ability to be reused. If you create a cable and it has tested poorly you can always redo

FIGURE 5-34 LMR600 Step 8.

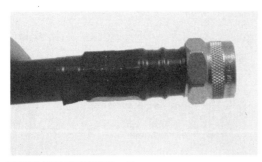

FIGURE 5-36 LMR600 using EZ-Connector Step 1.

FIGURE 5-35 LMR600 Step 9.

the connectors. The cable stripping for the EZ N-Type Connector is similar to using the crimp-on connector (Figure 5-30).

Once the cable is ready for the connector, slide on the first part of the connector as far down the cable as possible (Figure 5-36).

After the first portion of the connector is on, screw in the second part as tightly as possible and then tighten it more using two crescent wrenches. Pliers won't work because they slip, so use the crescent wrenches for safety (Figure 5-37).

Once the connector is secure, start the weatherizing process by wrapping the end with rubber or electrical tape.

Wireless Troubleshooting

Troubleshooting wireless networked video isn't much different from trouble-shooting an ordinary WLAN, with one exception: video surveillance streams in real time and demands better band-width. These issues aren't a problem with a WLAN providing e-mail and Internet access to numerous client

FIGURE 5-37 LMR600 using EZ-Connector Step 2: lock the two parts in place using crescent wrenches.

FIGURE 5-38 LMR600 using EZ-Connector Step 3.

FIGURE 5-39 LMR600 using EZ-Connector Step 4 – always weatherize.

stations, but it can become overwhelming when the WLAN is streaming video. Let's face it, if e-mail is down for an hour it may go unnoticed, but streaming video that goes down, if regularly monitored, will be noticed and is mission critical. If e-mail is down at the client station, those e-mails still reside on the server waiting to be downloaded to the client at the next interconnection. When cameras configured for video surveillance are down, they aren't recording so those archived data are lost forever.

Wireless troubleshooting becomes far more complex in a DVS system, especially if there's a mixture of different components. Troubleshooting a single wireless IP camera is one thing, but troubleshooting a mesh network of analog cameras connected to encoders, IP cameras connected to switches, and the radios involved begins with first determining if there's a wireless problem.

As mentioned in Chapters 3 and 4, the first course of business when troubleshooting, if you are unfamiliar with the system, is to create a data flowchart (see example in Figure 5-40). This chart provides a detailed view of how the video flows and exactly which devices it touches during its journey. This is important because the wireless aspect of the architecture creates another layer of complexity, but this isn't always the problem.

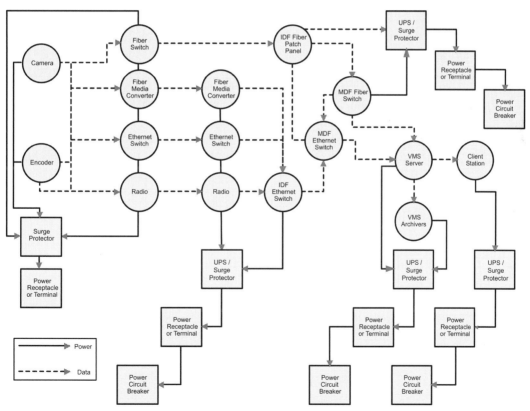

FIGURE 5-40 Sample data flowchart.

In fact, once any interference issues and channel planning have been addressed, the wireless layer is rarely the problem.

There are two basic problems that occur in a wireless DVS system: poor video and no video. Each one may have different issues, causes, repercussions, and solutions. Obviously, the problem of no video is more self-explanatory than poor video, which can be defined in a number of different ways. For the purpose of this troubleshooting section, the following symptoms are recognized as potential wireless issues:

1. Network latency
2. Artifacting
3. Stuttering while PTZ
4. Pixilation
5. Video blinks on and off
6. Waiting for Signal
7. No video

Network latency can be mistaken as artifacting or pixilation, but is mostly recognized when using a PTZ function on a camera. When panning, the image shows frozen images onscreen or delays panning during operation. Artifacting is when chunks of frozen squares are visible throughout the video image and pixilation is a more granular mosaic distortion. This differentiation occurs because artifacting and pixilation are caused by different issues.

Stuttering during PTZ causes the entire video image to stutter, as if multiple frames are dropped. Another issue is when the video blinks in and out, but this is different from when the encoder and/or camera icon in the VMS blinks in and out. When the video stream sporadically drops it becomes a communication issue between the VMS and the encoder and/or IP camera, but when the encoder or IP camera drops communications to the VMS a different set of protocols may have been affected.

The Waiting for Signal is defined as an encoder and/or IP camera linked to the VMS archiver software with no visible video image. This typically means that the client VMS software recognizes the device but not the video stream or method by which the video is streamed.

As mentioned in Chapter 3, the no video problem is depicted differently by VMS viewers and devices. It can be presented as simply a solid black image with the words No Video in the frame, a half red and half black frame, a solid blue frame, or with a No Camera icon in the frame. If any one of these warnings appears when an analog camera is connected to a digital video encoder, it typically means that the camera-to-encoder connection is interrupted. This is caused by the camera losing power, the coax cable melted by an enclosure heater (don't laugh, it happens), a bad coax cable, a poorly connected coax – such as when a BNC connector is set but not locked – or the encoder's video connection shuts down or fails.

■ ■ ■ ━━

Troubleshooting Tools and Equipment

Besides the required laptop computer (see the Troubleshooting Laptop section in Chapter 4), whenever and wherever troubleshooting takes place, it's best practice to have a known-good module of each component that needs to be tested. Typically that includes:

1. Encoder
2. Camera
   a. IP camera
   b. Analog camera
   c. PTZ printed circuit board
   d. Enclosure
3. Cables and wires
   a. Ethernet cable
   b. LMR antenna cable
   c. Radio to antenna dongles and/or whips
   d. Power extension cord
4. Ethernet switch
5. Wireless radio
6. Antenna (one of each used)
7. PDU
8. Remote power controller
9. Surge suppressor
10. Portable analog monitor
11. Patience

It can be difficult to determine which component is defective or has malfunctioned without a known-good replacement on hand to test it. Don't ever use these known-good components as replacement parts. It's easy to replace a problem with the closest one at hand, but the next time there's a call for field testing that component will be unavailable. Keep the test bed intact.

━━━ ■ ■ ■

One of the most important troubleshooting tools I've uncovered in the field is a wireless mobile unit. Most of my wireless networking involves mesh radios and a mesh radio can only talk to another mesh radio. Although you would troubleshoot a mesh network the same as any other wireless implementation, one of the major benefits of having a wireless mobile unit is the ability to configure this mobile radio to join the select suspect mesh radios. If the radio is out of reach, such as on a tall pole, configuring a mobile radio to join the mesh immediately determines if there's a wireless problem, a camera/encoder problem, or a power problem.

If the mobile radio joins the mesh network, the installed radios become visible so the power is on (radio status lights show if the unit has power, if it's successfully communicating, and if the radios are linked). At this point, there should be direct access available to the camera/encoder either via a Web interface or Telnet. Direct access proves there's nothing physically wrong with the camera/encoder and the problem is just the communication between the video streaming device and the VMS system or, in other words, a networking problem. This could be the Ethernet, between the portal radio and the VMS server, or from the mesh node radio to the portal radio.

Figure 5-41 is a flowchart depicting the troubleshooting steps for poor video. The complexity of these systems usually involves a variety of different devices, all separate, but made to work together. However, there are certain assumptions made when troubleshooting poor video as opposed to no video. Poor video indicates the problem is in the wireless connectivity. The signal may be poor, or there may be other factors within the system, but the bottom line is if there's video streaming from a camera connected to a wireless link, that radio is up and running on the correct channel. Remember that poor video means there's no need to troubleshoot wireless connectivity, just the quality of that connectivity.

Poor video could mean:

1. There's still power to the radio and the camera/encoder.
2. The radio is still on the correct frequency.
3. Waterlogged antenna cables and/or antenna connector.
4. Antenna is a directional antenna that has been misaligned by weather or vandalism.
5. Firmware incompatibility.
6. Firmware corruption.
7. Interference.
8. Wrong antenna (not in a compatible frequency).

Considering that all of these devices are run by firmware (embedded software), with the majority built on top of the open source Linux kernel, the first step is always a simple reboot of the suspect device.

Reboot

Know where the on/off switch is for the devices. If done properly, all devices in a single enclosure would be on a single circuit breaker. If that circuit breaker is inaccessible because of location, security, key management, etc., then integrate a remote power management unit (see Chapter 3). This remote reboot saves time and headaches, especially when under the gun (Figure 5-42).

When troubleshooting connectivity issues between wireless DVS and mesh radio/camera, first check active power and status lights. If all operational lights aren't on, reboot/recycle the radios. If all operational lights are on, make sure that the antenna

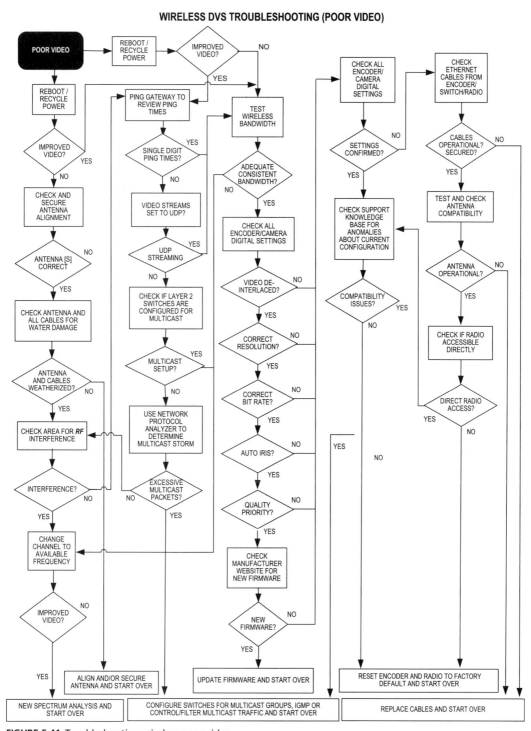

FIGURE 5-41 Troubleshooting wireless poor video.

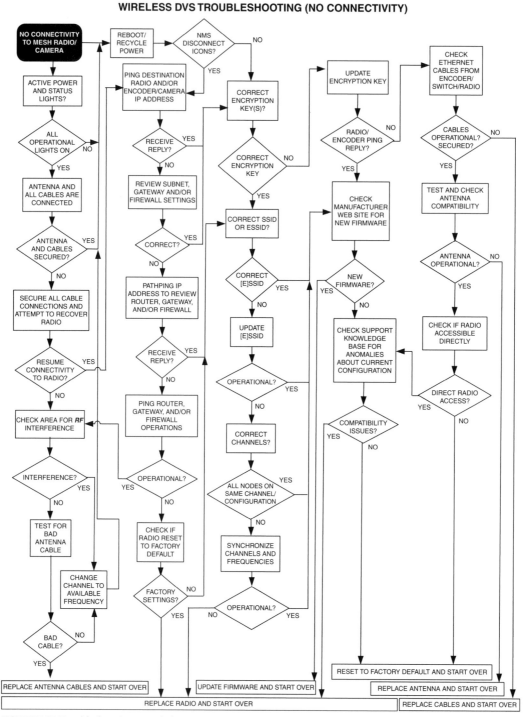

FIGURE 5-42 Troubleshooting no wireless connectivity.

and all cables are connected and secured. Once the antenna and all cables are secured, attempt to recover the radio. If connectivity is resumed, do a PING test to determine the quality of the connectivity. If connectivity isn't resumed, test for RF interference. If there's no interference, test for bad antenna cable. If your test reveals a bad cable, replace it and start from the beginning. If there's RF interference or if you find no bad cable, change the channel to an available frequency.

Reboot and recycle power and when the radios reappear in the NMS do a PING test to the digital video encoder or IP camera to determine the network connectivity. If you don't receive a reply, review the subnet gateway and and/or firewall settings. If the subnet gateway and or firewall settings are correct, PATHPING IP addresses to review the router, gateway, and/or firewall path. If you don't receive a reply, ping the router gateway and/or firewall IP addresses individually. If the router gateway and/or firewall are operational, go back and test for RF interference and continue from there.

If the router gateway and/or firewall aren't operational, check if resetting the radio to factory default and then reconfiguring will clear the problem. If the encryption key(s) is correct, make sure you have the correct SSID or ESSID. If the encryption key(s) is incorrect, update encryption IDs. Ping the radio encoder. If you don't get a ping reply, check Ethernet cables. If you get a reply, check the manufacturer's Web site for new firmware. If yes, update firmware and start over. If there's no new firmware, check support's knowledge base for anomalies about current configuration. If compatibility issues arise, replace radio and start over. If all nodes aren't on the same channel configuration, synchronize channels and frequencies and they should become operational; also check the manufacturer's Web site for new firmware. Check and make sure that all cables are operational and secured and if not, replace cables and start over. If all cables are operational and secured, check for antenna compatibility both in brand and frequency as well as polarization.

Chapter Lessons

- Always check for power at each location requiring a radio. Wireless doesn't necessarily mean completely without wires.
- Design for the right amount of coverage with a propagation tool if possible, and check the availability of the chosen RF frequencies through a spectrum analysis.
- Separate and secure wireless networks with encryption, strict ID and passwords, and if possible, hide the SSID name.
- Creating a wireless connection with a clear LOS is required for the maximum throughput and trouble-free Fresnel zone.
- Choose the right antenna for the requirements and, more important, the fluid RF environment.
- Uncover other users of the required frequencies and together develop a channel plan that shares the spectrum rather than competes in it.

Approaching the Project

6

Site Survey

Introduction

There are many types of digital video security (DVS) site surveys for camera locations, network equipment, conduit pathways, raceways, and ducts, both above- and belowground. The objective is gathering and then documenting findings and information to assist in determining if the desired location is suitable for the necessary equipment and meets the conceptual design requirements. This information may include:

- Area of coverage (exactly what is to be monitored)
- Power availability (existing exterior and interior power conduit and pathways for high, medium, and low voltage)
- Data pathways (exterior above- and belowground duct banks, raceways, etc.)
- Existing internal network topology (with existing data pathways)
- Existing network configuration, availability, and accessibility
- Existing main distribution facility (MDF), interim distribution facility (IDF), data center space, and accessibility

The area of coverage refers to the area where there's a need to monitor and/or archive activity. This could be a 360° area (back-to-back cameras on top of a pole), a single doorway, an outdoor terrace, or a parking lot. Wherever the location, it's assumed that someone, somewhere wants this location monitored and/or archived to present a capable guardian as a deterrent for any activity taking place as well as for video forensics. Understanding the purpose of and the reason for the video surveillance will assist in choosing the best location for the camera. For example, if the requirement for a camera is to monitor a parking lot for vandalism and theft, installing a camera that can be reached, either by climbing onto the roof of a vehicle or by ladder, only adds another potential target for an assailant. In this particular example, the higher or more out of reach the camera is, the safer it is from being rendered inoperative.

Other areas of coverage are unique by project and location but can typically include:

- Doorways
- Corridors
- Parking lots
- Common areas
- Public transportation
- Vehicles

- Sensitive locations
- Potential terrorist targets

Monitoring doorways offers a video record of those who exit and enter the facility and may show what they have in hand. Doorway points of interest may be something as simple as monitoring the front door of a home, anonymously reviewing visitors ringing the doorbell, monitoring loading docks, or recording vehicles (and license plates) moving in and out of a parking garage.

Adequately capturing the area of coverage for any doorway doesn't include mounting the camera anywhere above the door; doing so will guarantee capturing any and all activity in and out of the doorway, but will only give you a great shot of the top of everyone's head. Typically, if installing an outdoor camera, the ideal location for doorway coverage is face on, across the egress. This becomes difficult if the mounting location is on a neighbor's property or a utility-owned pole. Yet mounting the camera on the inside of the doorway leaves it vulnerable to vandals. Installing the camera on the same wall of the doorway, at least 25 feet opposite the door hinges, will give a good line of sight (LOS). If double doors are used, set up the camera in the LOS of the opened door, locking the other door to force individuals to exit facing one way.

Besides the obvious vandalism, drug trafficking, theft, and potential traps for other violent crimes, corridors are also monitored for liability reasons.

Determining the area of coverage also requires understanding the specifics of cameras chosen for the solution. If the surveillance requires monitoring during both day and night, then a low lux camera (see Chapter 3) would be required. If the area of coverage is pitch black at night with limited or no visibility, then an infrared (IR) camera may be a good option.

When choosing surveillance solutions consider the angle of coverage provided by each camera. If the parking lot is the required area of coverage to monitor traffic flow or potential theft and vandalism, then a fixed camera with a wide-angle lens may be enough. However, if license plate recognition (LPR) and human and/or face recognition are also requirements, then a different camera is needed.

Video surveillance is mostly used for scene overview – monitoring all activity within a specific location. The depth of detail required in scene overview guides camera choice – for example, if the requirement is object identification beyond the make and model of an automobile or identification of whether a human figure is male or female. Unless you're planning on covering 20 specific locations with 20 cameras set at a $10°$ area of coverage (with a telephoto lens), then pixel depth may be more important than the lens and angle or view.

A megapixel camera (see Chapter 3) is typically chosen because of a need for a larger area of coverage. The idea is that the camera provides more detail; therefore using a fisheye lens to monitor the entire area would satisfy the requirements. However, using a fisheye lens to cover as much of the area as possible doesn't necessarily give you more pixel depth, just more general coverage. Human face recognition and LPR both require more pixel depth than the average surveillance camera.

License Plate Recognition

LPR uses automatic number plate recognition from a video image of a license plate and then uses an optical character recognition (OCR) algorithm to convert it to the ASCII character set (see Chapter 3 for more details). For this to be a successful representation of the license plate number, there must be a minimum of 25 pixels on the vertical side of the license plate characters. That would mean that a total of 5000 pixels represents the entire license plate. Figure 6-1 shows the capture quality of LPR software.

The quality of the lens, the charge coupled device (CCD), complementary metal oxide semiconductor (CMOS) sensor, frame rate, and the OCR software are all components for successful LPR, but the number of actual pixels is the core of the conversion from blurry image to factual ASCII representation of the license number. A successful camera installation for LPR requires a camera focused on a position where license plates are at least 5000 pixels of the entire image.

Human Recognition

Facial recognition is much more complicated than LPR, as the human face moves far more fluidly and may include various accessories such as sunglasses, hats, and scarves. If the video capture of a face were to be straight-on to the camera lens, without any accessories, the bare minimum to identify a human face would be 25-75 pixels just between the eyes. This means that about 10,000-20,000 pixels are needed to identify an entire face.

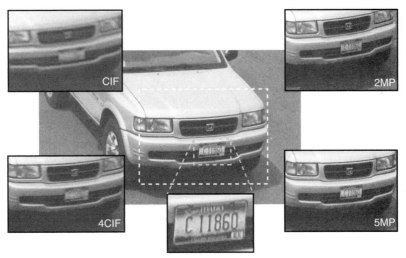

FIGURE 6-1 LPR using CIF, 4CIF, and megapixels (MP): LPR software is specifically designed to read license plates beyond the capability of the human eye.

Determining the distance from the camera to the object of interest requires both the horizontal distance from the camera and the camera's height. There's a significant difference between the horizontal distance to an object of interest and its actual distance once you've factored in the height of the camera. For example, if the goal is human facial recognition or identification, a camera on the side of a building may be 30 feet above a targeted egress. If a thief is 20 feet away from the egress, with the camera 30 feet above, the real distance is 36 feet, or 180% of the original horizontal distance.

Calculation of the true distance can be determined by using the Pythagorean Theorem ($A^2 + B^2 = C^2$). This determines the length of one side of a triangle based on the lengths of the other two sides. Figure 6-2 depicts the distance to the face of the figure as 12 feet. The camera is placed 6 feet above the figure's head. The exact distance is calculated as $6^2 + 12^2$, which equals 180. To determine the distance to the subject's face, you must find the square root of 180, which is about 13.5 feet.

The dimensions of the scene are then used to calculate the number of pixels representing the figure's face. A tool called a *lens calculator* can be used to determine the dimensions of the scene captured by a camera. If you Google lens calculator a free number comes up for you to use or you can choose the IP Video System Design Tool, a small undiscovered gem at www.jvsg.com (see Figure 6-3).

To use the lens calculator you'll need the focal length of the camera lens and the size of the CMOS or CCD sensor (see Chapter 3). Let's assume a 1/3" sensor and an 8 mm lens. The lens calculator determines the scene dimensions to be about 6 × 8 feet (a standard 4 × 3 aspect ratio), making the scene 48 square feet (6 × 8 = 48) at 13.5 feet away. An estimation of a human face at that distance would be about 12 × 6 inches, taking up about half of a square foot or 0.01 of the scene (1%).

The number of pixels representing the face will be calculated by determining the resolution used for monitoring (see Chapter 2). A single frame at CIF resolution (352 × 240) includes a total of 84,400 pixels. At 4CIF resolution (704 × 480) there are 337,920 pixels, 1,310,720 pixels in a 1.3-megapixel camera (1280 × 1024). A 3-megapixel camera includes 3,133,440 pixels per frame.

Figure 6-4 compares the number of facial pixels in three resolutions. The CIF image of the face would be represented by 775 pixels, while the 4CIF image would be 8680 pixels. The face on a 3-megapixel camera embodies 42,525 pixels.

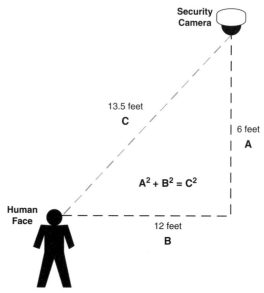

FIGURE 6-2 Use the Pythagorean Theorem to find the real distance of the human figure.

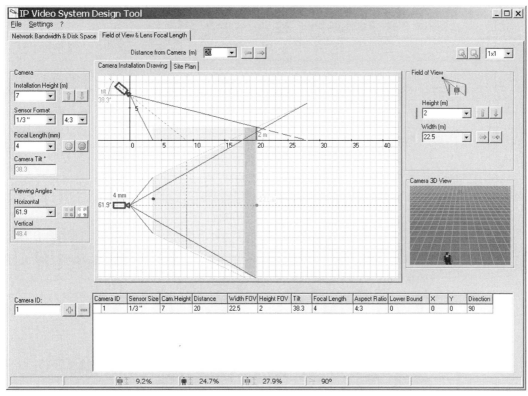

FIGURE 6-3 The IP Video System Design Tool, available at www.jvsg.com, provides excellent functionality for storage and complex field-of-view calculations.

See Table 6-1 for the suggested percentage of pixels required for facial and license plate recognition and Figure 6-5 for an example of facial images based on this table.

What I've used here is a simple example. However, in the outside world cameras are installed at least 25 feet high, away from unauthorized access and vandalism. When a camera is placed 25 feet high, if someone walks by at about 15 feet, you'll have a formula that

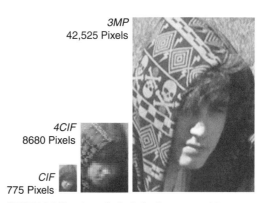

FIGURE 6-4 Number of pixels for face recognition.

looks more like $20^2 + 15^2 = ?^2$ or $400 + 225 = 625$. The distance between the camera and the subject's face would then be 25 feet. Using the same 1/3″ CMOS or CCD sensor and an 8 mm lens, the square footage would be 15×11 or 165 from 25 feet away.

Table 6-1 Percentage of Recommended Pixels per Frame for Facial and License Plate Recognition

Object	CIF	4CIF	1 Megapixels	3 Megapixels
Pixels per frame	84,400	337,920	1,310,720	3,133,440
Facial pixels percentage (20,000)	24%	0.06%	0.015%	0.0063%
License plate pixels percentage (5000)	0.06%	0.014%	0.0038%	0.0015%

Daily Storage, 30fps
~17 GB (MPEG4 - Average Quality)
~8 GB (H.264 - Average Quality)

Daily Storage, 30fps
~200 GB (MPEG4 - Average Quality)
~55 GB (H.264 - Average Quality)

FIGURE 6-5 Example of facial images based on Table 6-1.

The size of the human head from 25 feet away would be about half of the previous example, making it about 6 × 3 inches or about 0.005% of the overall area of coverage. At that size, with the same three previous resolutions (which wouldn't change), you have about as many pixels for facial and/or LPR. Table 6-2 summarizes the calculations used to determine number of facial pixels per resolution size.

Table 6-2 Number of Facial Pixels per Resolution Size

Camera ID	Height of Camera (A)	Horizontal Distance to Camera (B)	Pythagorean Theorem ($A^2 + B^2$ = C)	Actual Distance from Camera (C^2)	Sensor Size	Lens Focal Length	Square Footage Covered (Lens Calculator)	Facial Size (%)	CIF Pixels	4CIF Pixels	3 MP Pixels
1	6 ft.	12 ft.	$6^2 + 12^2 = 180$	13.5 ft.	1/3"	8 mm	48	0.01	775	8680	45,525

Power = Camera, No Power = No Camera

The initial site survey often serves merely as a preliminary introduction to the desired location, its target, and accessibility in the area. The site survey may only serve as a means for exploring the area. But remember that where there's no power, there's no camera, so determining power availability is essential on the first exploratory site survey. As part of the site survey, be prepared to explain if there's no cost-effective method to deliver power to the required location. Keep in mind that power doesn't necessarily need to come from the same location where the data are terminated, although having a separate circuit breaker for the cameras, with enough of a load to handle them, is always preferred. A single circuit breaker per camera is the optimal practice, because if the camera requires maintenance then only that single breaker shuts down the power for that individual camera and not a daisy-chained group (see Figure 6-6). Remember

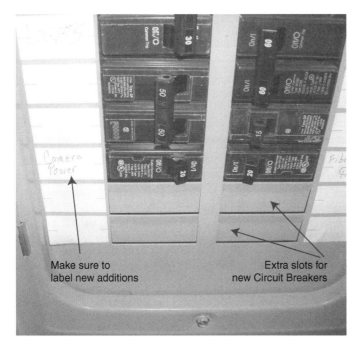

Make sure to label new additions

Extra slots for new Circuit Breakers

FIGURE 6-6 Check for space for dedicated circuit breakers for the cameras and label them appropriately.

if the cameras are down and there's a serious incident, there may be a legal liability for providing a false sense of security, so the fewer cameras down during maintenance, the better.

My work on Operation Virtual Shield (OVS) required supplying power to various city light and traffic poles. Most city and public area lights are higher voltage, and more important, are either set on timers and/or use photo sensors to determine natural light, which enables them to automatically turn on or off at dawn and dusk. The power requirement for all security cameras (e.g., fixed, analog, CCTV, IP, etc.) is considered "uninterrupted power" (i.e., AC power that's available 24/7 and is typically a clean run of 110/120 V). For the OVS project, every pole with a new camera required new individual runs of 12-gauge wire from the source as well as a single dedicated circuit breaker. The pathway ran under the street, through manholes, and up the selected pole. In contrast, Marquette Photo Supply (case discussed in Chapter 1) required only a simple installation of four fixed cameras. Power was run to the back office and the 12-V adapters were plugged into a single surge protector.

■ ■ ■ ▬▬▬▬▬▬▬▬▬▬▬▬▬▬▬▬▬▬▬▬▬▬▬▬▬▬▬▬▬▬▬▬▬▬▬

Feel the Power

Power and data should *never* run in the same conduit unless the power is considered LOW voltage! Power over Ethernet (PoE) will run power over pins 4 and 5 and 7 and 8 of the Cat5, 5e, and 6 cables using the standard RJ45 connector. The absolute maximum voltage allowed is 51 V (40 W), although most IP and CCTV security cameras can run on as little as 24 V. Never add data in a conduit with 110, 120, 208, 220, or higher voltage as the electricity will create havoc with the electrical pulses used for digital data delivery or the light pulse used for fiber (and may even melt the fiber if it's high enough voltage).

▬▬▬▬▬▬▬▬▬▬▬▬▬▬▬▬▬▬▬▬▬▬▬▬▬▬▬▬▬▬▬▬▬▬▬ ■ ■ ■

Surge Protectors and Suppressors

Voltage is a measure of the difference in electric potential energy; electric currents travel from point to point because there's greater electric potential energy on one end of the wire than there is on the other end. The principle of water pressure is an example of similar movement. Water, under pressure, will flow out of a hose because the higher pressure on one end of the hose pushes water out. Voltage is a measure of electrical pressure. A power surge, also called transient voltage, is a significant increase in voltage above the normal flow of electricity. Typical household and office wiring in the United States uses a standard 120 V. If the voltage rises above 120 V, it can destroy the equipment using that power.

When an increase in voltage lasts 3 ns or more, it's called a surge; if it lasts less than 3 ns, it's called a spike. Even if such surges and spikes don't immediately destroy equipment, they will wear the components down over time, burning up wiring and capacitors. Nothing speaks to the need for surge protectors and suppressors like a surge of power that fries all the equipment that took days to install. Many power supplies include built-in surge suppressors, so check equipment documentation and if surge protection isn't included, add to the bill of materials an appropriate model for the size of the system.

UPS

Uninterrupted power supply (UPS) continues to supply electric power to the required equipment for a select period of time when the traditional power supply is lost. Typically, the UPS is used for computers to prevent them from a hard shutdown that may damage the operating system and mission critical applications.

The UPS system standard IEC 62040-3 defines the amplitude and deviation of output voltage acceptable for switching the AC power supply to the DC battery backup. Since the UPS doesn't require electricity from an outside source to function, it uses a number of energy-storing backup batteries, an AC-DC charger to keep the battery fully charged, and a DC-AC inverter to provide the necessary power to the required equipment. The primary UPS type used for electronic surveillance systems is the Standby UPS, which will activate the battery backup when traditional AC input power fails. The typical power transfer time is between 2 and 10 ms, depending on the amount of time it takes to detect the lost utility voltage and turn on the DC-AC inverter. During this time the power to the load is momentarily interrupted. The equipment's power supply typically includes a ride-through or hold-up power stored within its capacitors. This hangs on until the UPS kicks into action, thus preventing data loss. The typical ride-through power of a PC has at least 15 ms.

Since the inverter must operate in standby mode, only operating when input power fails, the Standby UPS has the highest efficiency (95-97%) and reliability ratings. Although it's the least expensive type of UPS, it typically includes software that automatically creates a soft shutdown of your operating system and turns off your computer.

A line interactive UPS regulates the input AC voltage using a filter and transformer. This is a bidirectional inverter/charger that's always connected to the output of the UPS to keep the battery charged. When the input power fails, the transfer switch disconnects AC input and the battery/inverter provides output power. There's also an Online UPS that always delivers all or some of the AC power through its inverter, even under normal line conditions, so there's no transfer time at all – switching power over instantaneously.

Many commodity UPS manufacturers specify only a volt-ampere (VA) rating. A 500-VA UPS is typically only 60% of their VA rating. This is a throwback to old computer power supplies that had power factors of 0.6, so when selecting the size of the UPS be sure to accurately determine the exact net wattage. For instance, 300-W power consumption under the 0.6 power factor rating requires a minimum of 500 VA (300/0.6 = 500 VA).

Digital video surveillance typically requires UPS or even a backup generator, depending on the nature of the implementation, and is always noted in the initial statement of work and project scope.

Camera/Video Site Survey

The camera/video site survey process includes aboveground and belowground elements to determine the viability of the target location. There are also specific interdependent tasks requiring approval at various stages, prior to proceeding. The site survey location is determined by a number of factors, but first and foremost is acceptance by the customer (the person or entity paying for the project). A formal agreement on site survey location may be included within an existing statement of work or as within the scope of the development process. Either way, there are many factors to consider before choosing a final installation location for any equipment for the entire DVS system.

The customer should have final approval of the time and location of each specific installation. You may need to obtain approval from divisional or agency experts, and it may require a single or multiple signatures. This is done for the protection of all parties involved. It's highly inadvisable to move forward with any installation without this formal acceptance written into a site survey form deliverable. This form goes beyond a simple written statement of "we will install ten cameras in the parking lot." The form provides the specific details that, depending on the size and scope of the project, can assist in better control and management for the life of the project (see Chapter 9). The Project Management Institute (PMI) teaches that there's time for every stage and step in the PM process as defined by PMI. Any project, large or small, can spiral out of control. As much as 75% of all projects fail because the scope and statement of work lacked enough detail to provide a clear roadmap for all persons and/or companies involved. A site survey form can gather enough detail to clarify exactly where and when those ten cameras will be installed in the parking lot.

The final site survey form should include the following information:

- Area of coverage
- Photographs of each view
- Photographs of primary and secondary installation locations
- General lighting considerations
- Power and power pathway considerations
- Data and data pathway options (IDF and IDF to MDF)
- Potential obstructions
- Location on a high level map
- LOS options (for a potential wireless solution)
- Required approval signatures

A video surveillance system would also require more than a single form, depending on the complexity of the project. Detailed documentation is necessary for each camera location independently along with its required area of coverage, network topology (including IDF locations and the primary MDF), and location of the VMS server and storage.

A site survey form is created, in part, from the system and functional requirements listed within the statement of work and may include the following elements:

- List of survey team members
- General information on camera location
- Area of coverage
- Photographs of each view
- Photographs of primary and secondary installation locations
- General lighting considerations
- Power and power pathway considerations
- Data and data pathway options
- Potential obstructions
- Location on a high level map
- LOS options (for a potential wireless solution)
- Required signatures

List of Survey Teams

Surveys can include an officially designated surveyor from the system integrator, the construction contractor, and even the customer or management entity.

The system integrator's surveyor provides verification of site requirements and most likely is the chief architect of the system. The contractor surveyor evaluates equipment, installation probability, and power availability. The customer surveyor confirms the chosen locations and area of coverage. Confirmation of the locations and area of coverage is important, as they can change when moving from the drawing board (conceptual) to the physical location (actual), especially if the initial conceptual design was done using only a Google Earth map or similar tool. For example, primary and secondary locations for each camera installation may be included in the initial conceptual design, yet when the contractor sees the actual location he may find the area lacking electrical or data pathways to one or both locations. The customer surveyor can determine the importance of that specific location, whether it can be canceled, or if another optional location must be uncovered. Information on any optional locations must be welldocumented. It may be a specific hot spot where surveillance is crucial or there's no conduit pathway for data but there's power. The site survey form is where these new options are noted and a new plan proposed. This information may be used in a detailed design plan by converting wired locations into wireless locations and existing locations changed to new locations.

Camera Location

There's a difference between how a DVS system integrator and a customer may understand the statements: "I want to cover the intersection of Randolph and Franklin" or "I want a camera to cover the rear exit." Those are general descriptors that may have been added to the statement of work and likely made sense to the customer and salespeople, but those statements lack enough information to do anything but an initial conceptual design. A DVS system integrator needs to know: Where at Randolph and Franklin? Corner of the northeast building? Southwest traffic light pole? Which rear exit? Personnel or freight? The site survey form is used to gather the necessary details to make sure the initial fluidity of the requirements are solidified and concrete information is achieved. Depending on the complexity of the project, the site survey may actually entail a number of site surveys with each one needing the involvement of different experts. It would be beneficial to visit the requested sites and gather information for documentation in the site survey form to clarify during the walkthrough with the contractor and customer. A complete site survey form captures the desired data – including specific area of coverage, height restrictions, lighting, environmental factors, security – and documents how to access secured locations. The contractor surveyor provides access to the IDF, MDF, telecom closets, rooms, power cabinets, manhole covers, and underground pathways through provided documentation and assists in the verification of both power and data pathways.

Thus, a complete site survey form turns "Randolph and Franklin" into a more specific description such as "the northwest traffic pole (not the light pole) at the intersection of Randolph and Franklin, on the Randolph side." "Covering the rear exit" may instead be described as "on the building east wall, 15 feet over the freight door, centered between the freight and personnel exit doors." Now the site survey includes a specific pathway for power and data.

Figure 6-7 is an example of how to best monitor a proposed area of coverage. This is a simple parking lot with one 18-foot light pole with dual back-to-back halogen lamps.

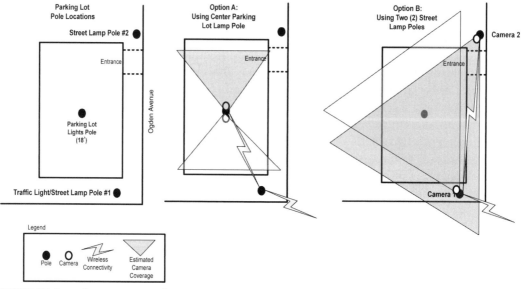

FIGURE 6-7 Area-of-coverage example.

The pole is mounted into the parking lot concrete. There's a traffic light on the southeast corner and a street lamp pole on the northeast corner, just north of the entrance/exit.

Option A shows the "usual course of action," which is to feed a new line of power to the center pole and mount two cameras, one pointing north and the other pointing south. In this case, high-end IP pan-tilt-zoom (PTZ) cameras were used and plugged via Ethernet directly into mesh radios, which then take the signal down the street to the office building and portal. However, most PTZ security cameras (IP or analog) have a field of view of about 50° in the wide-angle position (about 2° at full telephoto), so mounting them in the middle creates two blind spots unless the cameras are set to continuous panning and/or autotracking. If a PTZ camera is used, however, keep in mind that the motors inside all cameras are mechanical and thus prone to wear. Continuously panning a scene or autotracking all motion reduces the life expectancy of the camera. Even if we set aside the impossibility of feeding a new run of uninterrupted power to the pole (because the conduit underneath the concrete is damaged and blocked from the weight of the cars over the years), Option A still has only one camera to monitor the parking lot entrance/exit.

Option B, however, adds a second pole, creating a new design that has two cameras monitoring vehicles moving in and out of the parking lot without blind spots or excessive panning. This option also presents the possibility of using fixed megapixel cameras.

Network Infrastructure Site Survey

The following flow diagram (Figure 6-8) outlines the basic requirements of the network infrastructure survey procedures.

The purpose of the network infrastructure site survey is to analyze possible data routes (fiber and/or twisted-pair) from the camera locations to the IDFs and back to the primary MDF, where the servers and storage are located. This survey may also include details about interconnectivity of separate building locations on a campus environment or across streets and neighborhoods. This process begins with detailed plans, topologies, schematics, and diagrams of any and all common data and electrical pathways and duct banks. This first step is to provide identification of actual data pathways on campus. Once those pathways are identified and documented, the site survey team (more than one team for a large implementation using concurrent

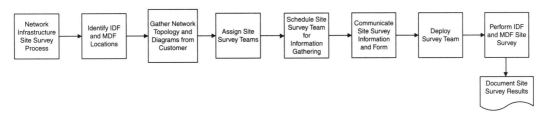

FIGURE 6-8 Network infrastructure survey process.

activities) analyzes maps and the physical underground routes to validate the findings. The teams will then map proposed data routes from the camera, IDF, and MDF locations. These proposed routes will be compiled in the network infrastructure site survey for approval.

■ ■ ■ ───

Cable Tagging

To clearly identify the cables during installation, the dedicated existing cabling, new cabling, and splices should be tagged by their specific functions. These tags should be photographed and added to the site survey form or installation documentation to clarify to the contractor which cables must be used for the implementation.

── ■ ■ ■

Interim Distribution Facility Survey

An IDF is a data intersection between camera locations and the primary server and storage data center, or MDF. The IDF can be a data closet, where a rack or cabinet resides for the Internet, telephone connections, or even part of an old CCTV installation (see Figure 6-9), and is usually environmentally friendly (e.g., air conditioned and dry). An IDF should never be located under large plumbing banks, basements, or attics – anywhere that can be flooded or can become extremely hot. The more networking equipment added to any IDF (or MDF) closet or room increases the room's temperature and heat dissipation must be considered.

FIGURE 6-9 CCTV installation requiring a DVS upgrade. Sometimes the biggest challenge is uncovering the cameras that are still operational.

■ ■ ■

Isolate the DVS Network

Either by virtual configurations (VLAN) or by physical hardware, it's best to isolate the DVS network from the rest of the business networks, because sharing bandwidth with sites like YouTube and other Internet trafficking may degrade how the DVS system functions. Don't inherit a poorly designed network, as it will negatively reflect on the DVS system.

■ ■ ■

The purpose of the IDF survey is to evaluate the customer-suggested site list for each IDF and to gain a thorough understanding of the physical aspects of the MDF and selected IDF sites from the standpoint of equipment installation and connectivity. The survey process starts with the identification of a potential IDF and the number of cameras that will terminate at that location. Considerations include geographic location, proximity to a range of cameras, accessibility and security, suitability for support of networking and networking equipment, and time to results.

Details from the site survey should always be documented in a network infrastructure site survey form in a standardized format using standardized terminology. The forms include a collection of logical network information (IP addresses, VLANS, etc.), physical location, rack diagrams, patch panels and cable entry/exit points, HVAC and power capacity, and contact and security details for future re-entry and service requirements.

Main Distribution Facility

The MDF is the heart of the DVS system, housing its server and storage equipment. It should be in a highly secure location, environmentally controlled, and include enough floor and/or rack space to hold the newly required equipment (see Figure 6-10).

If rack-mounted equipment is to be used, measure the height and width of the equipment rack and any available space that's required (e.g., 7', 19″ Chatsworth relay racks). Rack-mounted battery UPS backup systems can weigh in excess of 1000 pounds, so make sure the specifications of the rack equipment can handle the weight of the accompanying equipment.

■ ■ ■

Shipping and Receiving

Depending on the size of the DVS project, there may not only be wooden pallets of equipment delivered, but they may need to be delivered to multiple locations and can weigh (with commercial UPS battery backups) close to 1000 pounds. Make sure the destination is notified of the delivery and remember to record who signed for it and when it was received.

■ ■ ■

Power and Grounding

The required power is entirely based on the type and amount of equipment needed to accomplish the project objectives. Heavy equipment use in the MDF may require as

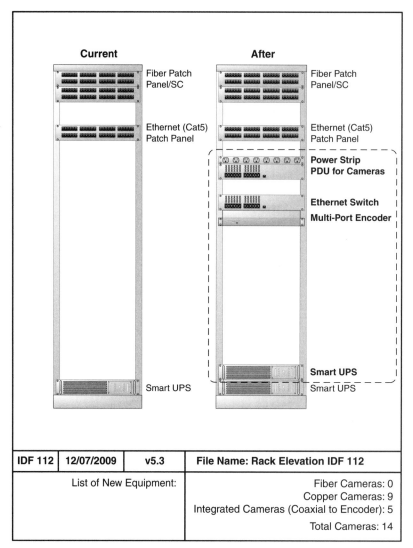

IDF 112	12/07/2009	v5.3	File Name: Rack Elevation IDF 112
List of New Equipment:			Fiber Cameras: 0 Copper Cameras: 9 Integrated Cameras (Coaxial to Encoder): 5 Total Cameras: 14

FIGURE 6-10 Rack diagram showing how the DVS implementation will change the IDF rack and/or cabinet.

much as two 30 AMP (amperage) – 208/220/240-V – circuits and one 20 AMP – 110/120-V – circuit terminating in a 1U rack-mounted 6-outlet power strip. Remember that whatever the specifications, you should install a dedicated circuit for the required equipment so you're not at the mercy of the existing electrical circuit.

"Grounding" or "to ground" electricity refers to the voltage level of "zero potential" through a "chassis ground" to the earth to safeguard the equipment from a build-up of static electricity or power surge. By grounding the power, the surge just trips the (dedicated) circuit breaker instead of frying all the equipment. A building ground to the relay rack should already exist if professional networking equipment exists within the IDF and MDF. The ground should be terminated on a copper ground bar and if there isn't one, don't inherit a potential disaster! Create your own copper ground bar.

Wireless Site Survey

As mentioned in Chapter 5, wireless doesn't mean there are no wires. The wireless site survey is an add-on to the previous site surveys to examine another option of delivering data, but there's still the need for power and connectivity back to the MDF. If there's available power to a required location, but data are a problem or too cost-prohibitive, wireless becomes a cost-effective solution.

Tools required for a wireless site survey include:

One spectrum analyzer
Two bucket trucks (optional)
Two laptop computers
Two wireless radios
Two 6′ coax jumpers
Two omni-directional antennas
Two panel antennas
Two AC power invertors

A wireless site survey can be broken down into smaller phases, but in the end the primary objective is to uncover all wireless activity in the area by using the required frequencies (channels).

Presurvey Exploration

This phase is strictly an information-gathering step to prepare for the physical wireless site survey. Information required for this stage includes:

1. Any documented use of required frequencies in the area of coverage
2. Farthest distance to be linked wirelessly (using Google Earth Pro)
3. Overall summary of terrain (using Google Earth Pro)
4. Required video quality, bandwidth, frames per second
5. Number of video streams
6. Days of archiving either continuous or motion detection

Once this information is gathered, it creates the baseline of power and wireless equipment required to achieve those objectives. Google Earth provides a means of visually documenting the data, terrain, distance, and LOS as well as taking note of other wireless access points (APs) and radios uncovered using the spectrum analyzer (Figure 6-11). Google Earth Pro offers a high-resolution printing feature for maximum detail.

Radio Frequency Spectrum Analysis

As explained in Chapter 5, interference from other wireless equipment is the biggest reason for wireless network failures and problems during either initial deployment or once the network has been established. Couple that with the constant demand for bandwidth-hungry, streaming video and I've discovered there's no such thing as too much information.

My current work is mostly in the public security 4.9-GHz frequency. That's a good thing because it's only 50 MHz in size and a bad thing because it's only 50 MHz in size. The benefit of a closed, licensed frequency is that it's relatively unused. Booting up a wireless-enabled laptop and scanning the available APs in a city the size of Chicago can bring up hundreds (of APs) in certain areas, depending on the power of the antenna. These APs are using the unlicensed frequencies such as 2.4 and 5.8 GHz.

Figure 6-12 shows a photo of my wireless mobile. I've set up a number of antennas, each with a purpose and stage of the overall spectrum analysis. The omni-directional, which reads 360° in the select frequency, shows if there's anyone using that frequency anywhere in the area. The sector antennas, either 90° or 45°, will help determine from what direction those signals are coming. Lastly, the powerful and laser-focused (10°) panel antenna pinpoints a more exact location. This helps in the overall design by selecting different channels and/or frequencies for the antennas pointing in that direction, or to cross-polarize the antenna to further protect the signal (see Chapter 5).

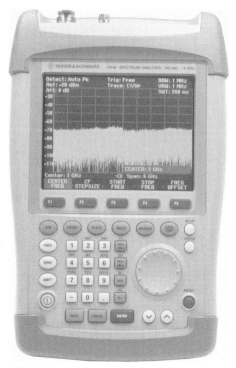

FIGURE 6-11 The Rode & Schwarz FSH6 Spectrum analyzer can detect any RF frequency up to 6 GHz.

Point-to-Point Mesh Radio Test

If heavy usage is required of frequencies, the best practice is to set up a live point-to-point (PtP) radio survey. By using actual radios with select antennas, a temporary PtP wireless connection is made to test actual wireless bandwidth, signal strength, RSSI, and video quality between the two units by using a select number of channels not in use during the spectrum analysis. The best way to do a PtP wireless test is to use two bucket trucks and raise the actual radios and antennas to the desired height. If the radios are to be installed between buildings in close proximity to each other and can serve as a radio frequency (RF) valley, then the bucket truck test is unnecessary. If the cameras and radios are to be mounted on the roof of these buildings, which are above the treetops, then the more information, the better. You may find that the required frequencies may be far too busy and other options need to be explored.

During the pre-survey stage, the mesh portal (or master) location is determined. This portal is the radio that links the wireless network into the wired network. That location can become the hub, with a number of select locations (all within the requirements for the area of coverage) including the second radio, your SUV, or on a truck. These tests are simple pass/fail with detailed documentation of how well each point passed in received signal strength indication (RSSI) and bandwidth.

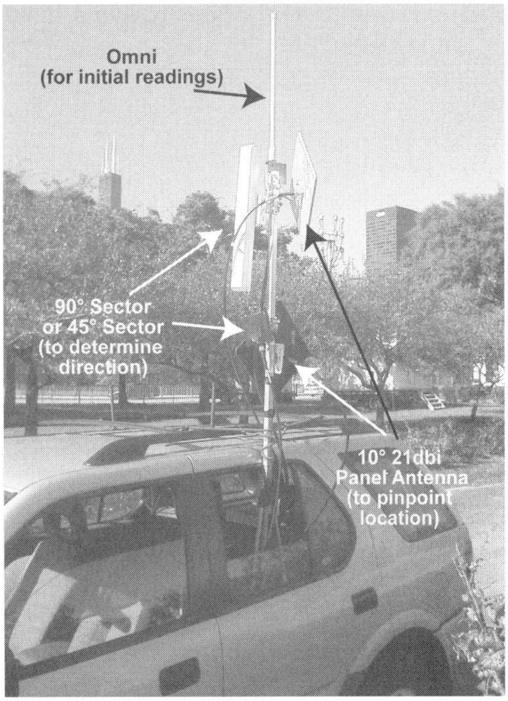

FIGURE 6-12 Wireless mobile has multiple antennas for specific functions all at an arm's reach from the spectrum analyzer.

Wireless Survey Results

The wireless collected data are additional to the official site survey form, because the point of the radio is to deliver video streams from the camera to a wired location, negating any installing, trenching, or boring of new conduit for data pathways or clearing existing blocked conduit pathways.

The following information should be added in detail for each radio location:

- Antenna type
- Antenna height (from ground level)
- Antenna orientation (if directional is used)
- Coaxial cable length from E-Box to antenna (if not mounted to box)
- Current radio signal-to-noise ratio (SNR) readings for each radio
- Spectrum analysis – note any possible interference
- Exact frequency and manufacturer's channel
- Signal level

Site Survey Tools

The successful completion of a site survey, whether wired or wireless, requires a select number of tools. Those tools can be hardware and software tools and be as complex as a laptop or as simple as a crowbar. Nevertheless, they're important to have on hand when visiting potential sites to avoid having to leave the site and return again with the required tool.

Site survey tools include:

- Laptop
  - Hardware:
    - Built-in Ethernet and wireless adapters
    - RS-232 serial port or USB with RS-232 adapter
    - USB ports
    - Dual processor
    - High-performance video graphics card
  - Software:
    - Windows with latest service pack, antivirus software
    - Turn off all firewall software
    - Video Management Software Client
    - Wireless Mesh Network Management Software
    - Microsoft Visio, Word, Excel
    - Adobe Acrobat
    - Adobe Photoshop
    - Google Earth Pro
    - Wireless propagation software
- Hardware tools
  - General tools (e.g., hammer, tape measure, pliers)
  - Crescent wrench
  - Screwdriver set
  - Nut drivers
  - Crowbar
  - Zip ties
  - Alligator clamps
- Digital camera
- Inter-team communications tools (cell phone)
- Spectrum analyzer

Laptop

The field laptop is a simple laptop running the latest version of Windows compatible with whatever software tools are required to accomplish the task. As depicted in Figure 6-13, there are notes and site survey forms to complete while on site that require a laptop for recording gathered information.

A built-in Ethernet and wireless network adapter is required for many site survey applications and becomes the hardware vehicle for gathering information on the wireless environment. However, using a separate PCMCIA card on your laptop (instead of using the built-in WiFi) isn't necessarily a bad choice (and a necessity if using a licensed spectrum), especially if you can attach a higher power antenna to it.

DVS Camera Site Survey Process

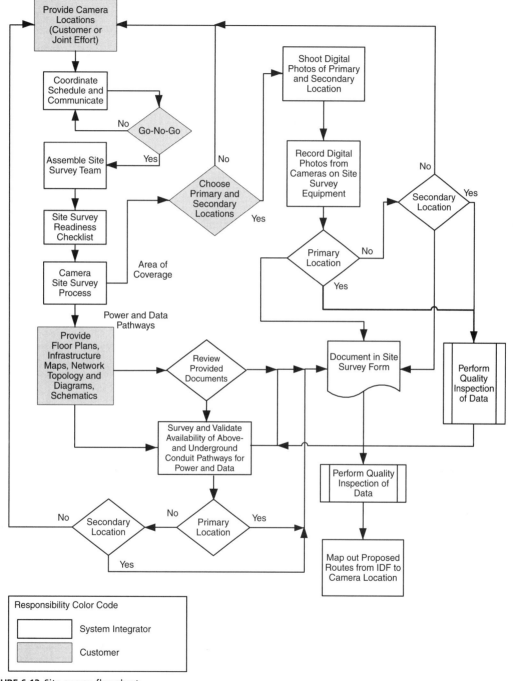

FIGURE 6-13 Site survey flowchart.

A field laptop also requires basic software necessities for documentation and deliverables, so an application suite such as Microsoft Office is required as well as a copy of Adobe Acrobat to convert those documents into non-editable deliverables (or just to shrink them down to e-mail size).

Adobe Photoshop is used to take screen captures and upsample them larger for print. A screen capture can be done in Windows by pressing and holding the CTRL key and then pressing the PRINT SCRN/SYSRQ key, just right of F12. This saves a copy of the screen on the clipboard. Create a new document in Photoshop and then press and hold the CTRL key again and then the V key to paste the screen capture into a new Photoshop image.

Google Earth Pro also provides a satellite view of the area of coverage as research for documentation purposes. Google Earth Pro also saves high-resolution files of the images, and this alone makes it worth the cost of the license. There are also wireless planning applications that use the information you input to provide an overall visual image of the area the wireless implementation will cover.

There are many developers of wireless propagation and/or spectrum analysis software, including companies such as Motorola (www.motorola.com/rfdesign) and Visi-Wave (www.visiwave.com; see Figure 6-14). There are basic link-planning applications

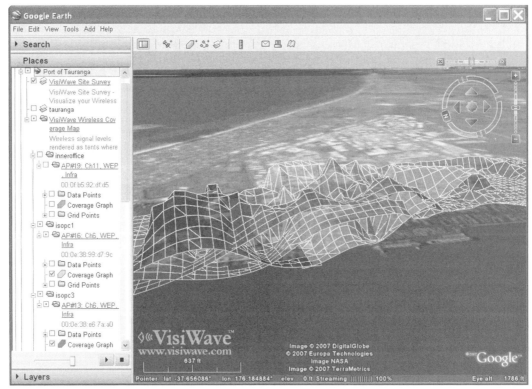

FIGURE 6-14 Wireless propagation software over an image of Google Earth Pro (photo used with permission by VisiWave and Google).

to assist in the proliferation of wireless APs to provide full coverage to a select area, but the more sophisticated (and expensive) applications offer more flexibility and reporting functions. This alone can save valuable time when creating a wireless site survey document from scratch.

Hardware Tools

When out doing a site survey, part of the objective is to explore. The reason for this is to uncover the best possible way of reaching the system requirements. How embarrassing would it be to design a solution that needs a thousand feet of new parallel conduit, when during the install a spare pair was uncovered? This is why you need your toolbox along with a few other select tools. As most of my site surveys are outdoors, I've found a crowbar to be a highly valuable asset. My reliance on these tools has grown to the point that I decided to buy a second set of tools (the ones I've become familiar with and accustomed to) just to keep in my vehicle. I've also found that temporary banding products such as zip ties in various sizes and industrial alligator clamps always come in handy when mounting antennas to record PtP RSSI signals.

Digital Camera

The site survey typically becomes a document deliverable and as such it requires details on the findings and recommendations. Thanks to digital photography, recording area of coverage, obstacles, discoveries, and specific camera installation locations is fast, cheap, and easy. A single photograph of an obstruction is worth a thousand words of description. Always bring a digital camera whenever doing a site survey.

■ ■ ■ ▬▬▬▬▬▬▬▬▬▬▬▬▬▬▬▬▬▬▬▬▬▬▬▬▬▬▬▬▬▬▬▬▬▬

Digital Camera Power

There you are with your customer, walking the area of coverage where he points out those unique areas that he needs monitored to improve the security and safety of the facility(s). These are very select locations and there are many. Good thing you have your digital camera to record these nuances in the customer's system requirements, but alas, upon pulling out your fancy digital pocket camera, you discover the rechargeable battery is dead. This is why I always carry a spare high-quality digital pocket camera that uses AA batteries and extra batteries.

▬▬▬▬▬▬▬▬▬▬▬▬▬▬▬▬▬▬▬▬▬▬▬▬▬▬▬▬▬▬▬▬▬ ■ ■ ■

Communications

Depending entirely on the size of the implementation, there may be more than one active site survey team. This is when interpersonal communication is crucial to share important information. Such communication can be accomplished (as usual) using a cell phone and/or walkie-talkie, but first the exchange of contact information is important not only to the discovery, but also for later documentation.

Chapter Lessons

Some things to remember when doing a site survey:

- It's an exploration of discovery with the prime objective of meeting the system requirements the best and most cost-effective way.
- Make sure the client is clear and specific about what the area of coverage entails; vague language can be confusing.
- Understand the equipment you're using and what it can or cannot accomplish so that you choose the right equipment for the job and place it correctly.
- There are particular camera, data, and power and network site surveys required as a bare minimum for a DVS system.
- The more details gathered, the more useful the information, so fill out a complete site survey form for each camera location.

7

Choosing the Right Software

Video Management System Software

There are many digital video management system applications available. The key to choosing the right application rests entirely on the system requirements, so the more detail included in the project requirements, the easier it is to make a decision. The many choices in the market (far too many to even consider exploring here) create serious challenges, as organizations sift through all the available VMS options. In a business environment, making a wrong software decision can increase organizational risk, cause inefficiencies, and eventually require costly upgrades.

Video management system software are constantly evolving to keep up with market trends and demands. Originally developed as an interface to monitor video surveillance camera streams, the convergence of telecommunications, media, and security provides a new proficiency that not only protects business and property, but people's lives as well. Newer functionality such as video analytics, people counting, autotracking, and alarm management further validates this demand. Avoiding pitfalls must be accomplished through informed decision making, which includes understanding not only the VMS software, its features, and functionality, but also the software company's market segmentation, distribution, and integration processes.

Software Applications

Behind the scenes of any company is the software they use, referred to as an application or a system. These applications make it possible to improve business processes and increase productivity and efficiencies. A good example is this word processor I am using, as opposed to a typewriter, which doesn't provide options such as cut-and-paste, delete, and undo or have a built-in dictionary and thesaurus.

Applications are now a necessity for making any business thrive, and like all software they need hardware to operate. Some applications require installation on top of other host applications, such as an operating system (OS). These are applications developed to be an interface for better business solutions. Applications are also developed to be embedded within a hardware chip to perform a single function, such as converting analog video into digital video like a digital video encoder.

Digital Video Appliance

The typical digital video recorder (DVR; or networked video recorder, NVR) includes a host application (usually a license-free Linux OS), video conversion software, storage and disc management application, and a VMS interface to manage the integrated video. The embedded video surveillance application and OS provide an appliance rather than a computer, streamlining the process of integration, maintenance, and support. The DVR was the digital appliance that replaced the analog videocassette recorder (VCR).

VMS software providers such as Genetec, OnSSI, Milestone, and ipConfigure are the revolutionary players for digital video security, providing flexibility as a software suite for a host application or as an NVR hardware appliance. These two solutions service different markets and are also driven by the size of the implementation. For example, Genetec, the leader in enterprise-wide VMS software, recently introduced the Smart Vault, a rack-mounted 1U server with up to 3 TB of storage space, and a Genetec Security Center license for up to 32 IP cameras, or up to 16 analog cameras through its built-in digital video encoders and 16 IP cameras. This packaged solution has an MSRP starting at about $6000, making it a viable option for smaller implementations.

The Genetec Security Center includes the Omnicast video surveillance VMS software, the Synergis Access Control software, and the AutoVU license plate recognition software for use with only those Smart Vault 32 cameras. If any additional cameras are added, the user will need to install another Smart Vault, configuring the two to communicate over the network. If the maximum possible cameras at any one location don't surpass the 32-camera limit, then this video surveillance appliance (VSA) is beneficial, but if there's any possibility of expansion, installing the Genetec Security Center onto an enterprise Windows server would provide unlimited camera licenses (with the proper archiving load balance). Remember, just because you can have licensing to archive 1000 cameras doesn't mean that the server can handle that kind of workload. Check each VMS system requirement for optimal performance limits.

The ipConfigure VSA is also designed for small to medium IP video surveillance applications and comes preloaded with ipConfigure's software in either 2- or 4-TB configurations in the same 1U server chassis. Providing MPEG-4/H.264 support for up to 32 IP cameras (at 10 frames per second; fps), it's shipped in either a RAID 0 (2- or 4-TB striped volumes) for maximum video storage or RAID 1 (1- or 2-TB mirrored volumes) for fail-safe redundant video storage. The ipConfigure software is pre-installed, reducing configuration and implementation time. The Windows Server 2003 and database (SQL Express) licenses are also included and preloaded and have an MSRP of about $6000 for the 4-TB storage version and the ipConfigure license (Figure 7-1).

There are many different solutions in the video alliance market, most limited to about 30 cameras. The Pivot3 solution offers greater expandability for storage using a RAIGE array, which is a collection of databank appliances connected with standard gigabit Ethernet connections where each databank runs the RAIGE storage software.

Databanks are automatically discovered on the Ethernet network and assigned to virtual RAIGE arrays of up to eight databanks.

The Intransa VideoAppliance comes in various sizes based on camera implementation of up to 70 IP cameras. It's also certified to use a number of different VMS software solutions, including Genetec, Milestone, and OnSSI.

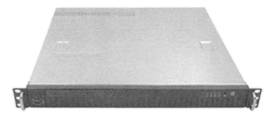

FIGURE 7-1 The DVA is an all-inclusive rack-mounted server with full licenses on professional VMS software for about 30 cameras.

Many of the VMS software companies also offer an appliance solution to simplify and expedite design and installation. The most costly segment of a digital video security (DVS) implementation is the installation process, especially if hundreds of feet of new conduit are required.

Processing Power

Video processing and recording demand heavy horsepower and again, depending on the size of the DVS implementation, it's best to not only have the archivers on a separate machine from the VMS server application, but the database as well. You need to give the VMS application enough horsepower to perform its functions without being bogged down by a continuously spinning hard drive recording video. Although a few software specifications consider 2 GB of memory acceptable for implementations under 25-30 cameras, the cost of RAM has dropped so dramatically that upgrading to 4 GB (the maximum Windows recognizes in the 32-bit architecture) can only benefit the performance. The simple rule for processing power specifications is any server archiving up to 30 cameras simultaneously should have a fast dual core processor. An archiver with 30-50 cameras should be using a quad core, and anything above 50 cameras should be using a dual quad core machine (a total of eight processors). Dual (2) and quad (4) core processors include multiple microchips to better perform multiple simultaneous functions.

Business Process Improvement

The addition of a new VMS software or solution into the business isn't like a typical software install. Traditionally, an IT department may install a new application for the employees to use and confirm their attendance for training. The employees attend the training, pick up enough to navigate around the application, and then truly learn by using it day in and day out. Successful VMS implementations, however, are based upon how well the integrator understands the software, the industry, and the internal business processes.

Figures 7-2 and 7-3 show before and after images of Chicago's Navy Pier DVS conversion. Figure 7-2 shows the Navy Pier "before," with a mixture of four disparate CCTV

FIGURE 7-2 Before the DVS convergence there were many separate video surveillance and alarm systems running on multiple interfaces *(photo courtesy of IBM and Navy Pier).*

FIGURE 7-3 After the DVS convergence there was only a single professional VMS interface *(photo courtesy of IBM and Navy Pier).*

systems. With the exception of the 16 food court IP cameras (pictured on the monitor top center), all of the other cameras were displayed on analog monitors by manually scrolling through each camera and/or watching them in quad frames on the analog monitor through a video matrix. There were no archiving and no video messages, and the panic alarms in the garages (which included high-pitch sirens), as well as the impact alarms in the Smith Museum of Stained Glass, triggered the closest camera to the location onto a single monitor. A panic alarm reset button was installed to turn off the siren once a risk assessment had been determined and the appropriate response taken. An incident report was filed for the archives at yet another computer within the security office. Although all cameras included abbreviated labels on the bottom of the screen, their exact location was unclear unless you were completely familiar with the 3/4 mile tourist and convention attraction.

This was a simplified summary of the business process for security. For those who have experience with DVS conversions and/or implementations, the quantum leap into the digital world would require a broader business process improvement program within any organization. This is important to create the most value from the VMS application. Simply dropping a new software application into an environment without training, guidance, mentoring, and documentation only creates confusion and becomes an obstacle for business process improvement.

Best practice dictates understanding and documenting the functions and outputs of the VMS software for each organization. This practice provides organizations with the knowledge and ability to develop key performance indicators, track the various measurable outputs, and enable better management of the solution.

Figures 7-4 and 7-5 depict the change in the single monitor of the CCTV video surveillance system and in the new digital interface of the professional VMS software, including onscreen pan-tilt-zoom (PTZ) controls, alerts, messages, a map of the selected area of coverage, and the specific cameras required for monitoring at that time by simply double-clicking the camera icons on the map. This opens up a specific camera within an open tile around the map, and the number of tiles (and cameras) is only limited to the workstation's horsepower and network's bandwidth.

The right VMS software, which meets business requirements, can and has been proven to improve business processes and produce measurable results.

Compatibility

All VMS software doesn't support all digital video encoders and/or IP cameras. There are also more granular compatibility issues, which is successful in retrieving video streams, but with a loss of functionality. Features such as H.264 compression or in some cases even regular MPEG-4-, relying only on the MJPEG stream or motion detection,

FIGURE 7-4 Typical CCTV interface, scrolling through cameras, one at a time.

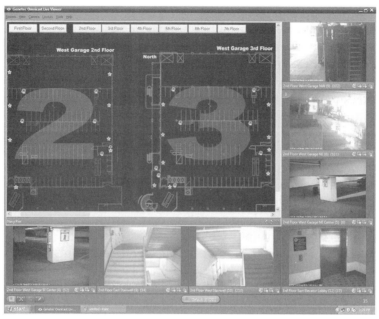

FIGURE 7-5 New VMS interface with map.

active tampering, or autotracking all become useless outside the Web interface of the device itself. The very basic requirement for any VMS software is compatibility, which allows control over the installed digital video encoders and IP cameras. This is accomplished using a unique software driver that becomes the interface between the OS, the VMS software, and the networked device. The more economical VMS solutions may have limited compatibility with only a select OS, because the application wasn't designed for enterprise-wide solutions or was designed to be used with select camera/digital video encoder manufacturers. Thus, it may only be compatible with the workstation OS and not a network operating system (NOS). Even the economical four-port DVR cards are developed with software that operates only with select cameras.

The software driver interface between the embedded hardware firmware and the VMS system (and the host OS) makes the video viewable and the controlling functions possible. Each hardware device may use a proprietary video codec (video compression/decompression), but without a driver and codec to recognize that stream, it can't display it or record it.

Any VMS software needs to meet the project's requirements. That's easily done when the project is new and the equipment adheres to the VMS system. However, if it's necessary to integrate existing assets or other security components, or an enterprise-wide solution with scalability, the choices become limited.

The purpose of the VMS software is to provide a single interface for the video surveillance solution, no matter what end devices are used. Depending on the size of the solution, it may be more cost-effective to replace all the legacy cameras rather than pay for an exorbitant license for a VMS suite that offers more than is required. Each professional VMS solution provides a listing of supported devices on their respective Web sites.

Market Share

The VMS market is constantly changing as new software-only vendors introduce solutions and upgraded versions. A recent study by IMS Research on the market share of the top VMS software packages may be skewed because the percentages used are based on revenues and not on units or licenses sold. Genetec Omnicast (www.genetec.com), Milestone (www.milestonesys.com), and On-Net Surveillance Systems (www.OnSSI .com) all continue to battle for the top position in the VMS open platform IP video management software field. All three VMS solutions include long lists of supported devices; from Arecont to Verint. These three are considered the top tier of VMS solutions with double-digit market share. Milestone Systems, with corporate headquarters in Denmark, battles for domination of the professional VMS software space with Genetec Omnicast, out of Canada, and On-Net Surveillance Systems, Inc. (OnSSI) located in Pearl River, NY. The professional VMS software suites include a well-documented Software Development Kit, which assists in the integration and seamless interoperability with access control, point-of-sales, analytics, biometrics, lighting or gate controls, alarms, fire alerts, baggage claims, and production systems.

The middle tier includes such applications as ipConfigure (see Figure 7-6), Video Protein, Lenel, and Exacq, which are all very capable with many features but also limited in supporting devices (see Table 7-1).

Metadata and the Database

A database is a software application designed to efficiently accumulate and store information. These data can be quickly retrieved and can have multiple relationships with other data within that database or other databases. This is an important component of any professional VMS software solution, where information (data) is used to store and retrieve archived video footage. Not only can a database-driven software retrieve specific information (e.g., the video surveillance footage at 3:10 a.m. on Wednesday), but the VMS software can also include what's called metadata. Metadata is simply data about data. For instance, a VMS software package may be configured to record all activity 24/7, but depending on its compatibility with the camera or digital video encoder, if you continue to monitor motion detection it will record that information as metadata. Thus, you can perform a search such as "any motion detected between 2:30 and 3:30 a.m. on Wednesday" and that exact footage will be retrieved, potentially saving hours of research.

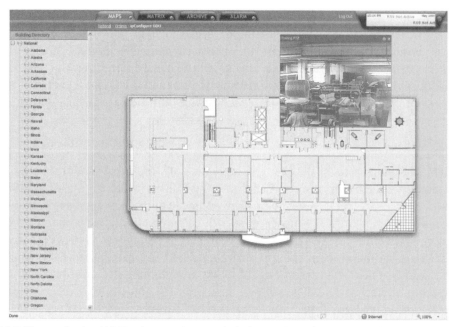

FIGURE 7-6 Most professional VMS software suites now include mapping solutions. IpConfigure also offers a mapping solution with live streaming video as a pop-up when scrolling over a camera icon.

Table 7-1 Applications and Supported Devices

	Cisco	Exacq	Genetec	Milestone	OnSSI[a]	ipConfigure	Video Protein	Lensec
Centralized Metadata	X		X	X	X	X	X	X
Standalone	X	X	X	X	X	X	X	X
Microsoft SQL Express			X	X		X		X
Microsoft SQL Server full version			X	X	X	X		X
Oracle	X					X		
My SQL	X	X				X		
DB2	X					X		
Windows		X	X	X	X	X		
Linux	X	X				X	X	
Sun Solaris						X		

[a]Customized database.

Video Analytics

The core foundation of DVS is the convergence of digital technologies and its software interface. Thanks to that software interface, DVS can become more intelligent simply by programming it to be smarter or adding programs to make it smarter. In Chapters 2 and 3, I analyzed facial and license plate recognition software, but those applications are only the tip of the iceberg. Even if the current VMS system doesn't provide better video analytical intelligence, there are new software applications being developed that use existing video footage to improve business processes and protect people and property.

In the world of video surveillance, the process of monitoring video onscreen for any unauthorized actions can be quite a daunting task when you're confronted with dozens of monitors. The odds of security personnel catching the right moment on the right monitor at the right time requires too many security personnel and monitors and too much time. VMS software can make this a simple task by doing the majority of the work through feature-rich security cameras, digital video encoders, and archiving the video for video forensics. Even if a crucial moment was missed on the live monitor, the archived video may have a record of it. Through metadata, the VMS software knows where that specific time and date is within the hundreds of hours of video footage sitting on the hard drives. All of the recorded video footage is bookmarked with times and dates for easy retrieval at a later date.

Metadata is also at the core of any video analytics software solution. For example, IBM's S3 (Smart Surveillance Solution) video analytics software requires about 5 GB of metadata alone, per camera per day. This metadata is a record of the software's memory, as it's programmed to do. This S3 software was developed to be smarter about what it saw visually within the video stream. The smart surveillance technology base includes a more sophisticated form of object detection, even in the presence of distracting motion, such as wind-blown trees, flags, and clouds. The two-dimensional auto-object tracking can continue to track one or more objects concurrently, once detected, thus providing independent object classification. Multiscale autotracking uses a PTZ camera to follow moving objects that qualify within a certain classification and can hand off the automated tracking across other cameras. S3 can also catalog faces from large distances, providing a database for facial recognition. The metadata is in XML format, representing a multitude of object and motion attributes. Real-time event indexing is instantaneously available for searching in a distributed and scalable database environment. All this is accessible through a Web service interface that also supports rapid application development (RAD) of customer-specific applications.

Besides the ever-growing list of capabilities built into the VMS software releases, other powerful video analytics applications include AgentVI (www.agentvi.com), which was used at Chicago's Navy Pier for virtual tripwires. Other solutions include MATE Intelligent Video (www.mate.co.il), Aimetis (www.aimetis.com), and March Networks (www.marchnetworks.com).

The following bullet points provide a comprehensive feature list of video analytics. Many of these features are embedded into the hardware firmware or can be triggered by analytics video management software suites. Each software vendor may not support the entire list, but that's only relevant if the system requirements demand that the analytics software do it all. The key factor is that the software and the hardware it's trying to communicate with and/or control are compatible:

- Real-time data retrieval
- PTZ autotracking
- Multicamera autotracking

- VMS integration
- Occupancy detection
- Area obstruction detection
- Intelligent scene verification
- People counting
- Perimeter protection
- Loitering detection
- Facial detection
- Queue length monitoring
- Camera tampering alarm
- Camera alignment detection
- Fall detection
- Traffic flow monitoring
- Panic alarm
- Abandoned object detection
- Speed monitoring

Real-Time Data Retrieval

The concept behind real time is one of instant access to data as it happens. Real-time data retrieval provides the ability to retrieve information, which may include video data and/or the metadata programmed for examination, as soon as it happens. These data must be available when most valuable; otherwise the system provides less value for security.

PTZ Autotracking

Once a moving target is detected, for example, a slow-moving vehicle entering a construction site after hours, the autotracking function can force the PTZ camera to follow that vehicle wherever it goes as long as it's within the camera's area of coverage. Multi-camera autotracking, on the other hand, provides that same function from the VMS software, which recognizes and controls the cameras and their autotracking abilities. The VMS has the ability to cross multiple PTZ cameras to continue to follow a target, so it can jump from one PTZ camera to another once the target is out of range.

VMS Integration

Video analytics also adds intelligence to existing VMS software beyond its typical features. The integration of alarm systems and electronic access control systems may require an additional functional interface to translate communications between those systems and the existing or new VMS software.

Occupancy Detection

There are occupancy detection sensors specifically designed to recognize the existence of humans within a confined space and count them from any angle. These devices, such as Lyrtech's Intelligent Occupancy Sensor, recognize humans by vision, not by motion. Some video analytics programs determine how many persons occupy a given space by noting the various moving objects, which isn't as accurate as recognizing those moving objects as human. Social gatherings in unique environments may include decorative elements such as balloons or flora, all of which may be misinterpreted as an occupant by simple motion detection. The better video analytics solutions can determine occupancy with non-motion, thus a person who remains motionless is still recognized as present. This avoids the common problem witnessed by people using motion detection light switches, which turn off lights in a room when the person is still, and that person must subsequently wave her arms to turn the lights on again.

Area Obstruction Detection
Video analytics can be programmed to recognize an open area within the frame and when that area is blocked for an extended period of time. The software is programmed to visually understand what it's monitoring and when that area is obstructed.

Intelligent Scene Verification
Similar to active tampering, intelligent scene verification recognizes when the camera field of view is altered or obstructed for any given length of time.

People Counting
Much like occupancy protection, people-counting applications must be able to recognize people visually and then count how many enter or exit the facility. Simple motion detection may misread objects as additional people.

Perimeter Protection
The value of perimeter protection is emphasized as far more than just motion detection when you consider the environmental factors around the area in question. For example, at Chicago's Navy Pier, standard motion detection won't achieve perimeter protection over Lake Michigan as the movement of the waves would hide the real target, rendering the application useless. In this instance it's important to use video analytics intelligence to create a virtual tripwire within a field of view to protect the perimeter. If any boat crosses over the tripwire, the video analytics software sends an alert to the VMS software and pop up a tile showing live video coverage of whatever boat crossed the line and show even prerecorded video up to the moment of activation.

Loitering Detection
The more sophisticated video analytics applications can differentiate individual pedestrians within the area of coverage and track them individually by appearance. This tracker feature identifies pedestrians according to the color of their clothing among consecutive frames. Each pedestrian is recorded as metadata in a visitor log with a time stamp to determine movement history, so even if the pedestrian moves out of the camera view, the analytics metadata can re-identify him upon his return to the area of coverage.

Facial Detection

There are three methods by which video analytics software can identify real-time face detection within color images. The software that uses more than one method achieves the greatest accuracy, but the application's purpose isn't to identify an individual, but only to recognize that there's a human face to capture. That's the difference between "facial detection" and "facial recognition." Facial detection software identifies a human face and its location, but the security personnel must identify the target. The first method used by video analytics software detects the human head through an outline analysis. The second uses the color of skin to first determine that it's a human, then the location of the face. The third method detects facial patterns, such as two eyes, two ears, hair or facial hair, a nose, and mouth in the right proportions. Obviously, indoor facial detection works better than having to decipher facial patterns through hats, sunglasses, and winter wear.

Queue Length Monitoring

Queue length monitoring is a way to automate the processes of controlling the number of people waiting in line. If there's a predetermined threshold, either for security or city code reasons, the software sends an alert to the VMS system along with a priority placement of any related video.

Camera Tampering Alarm

While intelligent scene verification monitors the actual scene, camera tampering monitors the camera and lens for obstacles and obstructions. Spray paint and/or paintball guns can render a camera useless without security acknowledging it, unless there's an alarm set to warn of wrongdoing. Once the system recognizes vandalism, the video analytics can turn any other PTZ cameras in the vicinity toward that location for observation. Motion detection can then activate an alarm within the VMS software to notify security of a potential culprit.

Camera Alignment Detection

When using a fixed camera for surveillance, that camera is mounted to monitor a select area of coverage. Intelligent software can determine if that area of coverage has changed because the camera has been physically moved out of its required field of view.

Fall Detection

Video analytics software is programmed to not only recognize an individual object as human, but also if that human takes a fall and records the incident for future video forensics review. How and why that person falls can be a deciding factor in a liability suit.

Traffic Flow Monitoring

In some situations, how the crowd is moving and how many move by during an allotted time frame, either in an open area or queue, can be as important as how many people are in the field of view.

Panic Alarm

While monitoring the traffic flow of people, an incident that causes a number of people to suddenly turn around and run in the opposite direction can be recognized by video analytics software as panic, setting off a number of preprogrammed alerts within the VMS software to present messages, priority positioning for the cameras in that select area, and even inform security personnel on the premises.

Abandoned Object Detection

In a world where terrorism is a reality, the threat of an abandoned object, which may be a bomb, can't be ignored. Video analytics extends security by turning all cameras designated for the area of coverage into trained observers. When any object has been left at a busy scene for a select period of time, an alarm is sent to security personnel through the VMS interface and even remotely to mobile devices. This key video analytics feature aids in the timely management of potentially dangerous situations. Abandoned objects detection has also been used to monitor illegal parking and to search the metadata in the archives for select programmed events such as blocked walkways or handicapped parking zones.

Speed Monitoring

Whether the camera is recording at 30 or 6 fps, if a person or vehicle moves through a field of view faster than a set programmed speed, the program can trigger an alert to security personnel, who can easily follow the tracks of an alarm and help decipher when and where a culprit may have run.

Salient Motion Detection

Video analytics is sophisticated software programmed to work like the human brain. It needs to recognize the difference between interesting (salient) motion (e.g., a person or car) and uninteresting motion (e.g., swaying branches or clouds). In the IEEE and IBM white paper titled "Robust Salient Motion Detection with Complex Background for Real-Time Video Surveillance" (*IEEE Computer Society,* January 2007), authors and researchers explain the new real-time algorithm that detects salient motion in complex environments by combining temporal difference imaging and a temporal filtered motion field. They discovered that the more interesting or important objects move in a consistent direction for a period of time and any previous information about that object, such as size and shape, is unimportant. Many video analytics software products, in an effort to recognize the more important motion within the video frame, attempt to remove the background through an analysis of hundreds of images, thus bringing any new salient motion to the foreground.

The new algorithm used in S3 effectively recognizes salient motion within a variety of real environments with distracting motions such as lighting changes, swaying branches, rippling water, waterfall, and fountains.

Object Classification

Moving foreground objects can be classified into relevant categories. Statistics about the appearance, shape, and motion of moving objects can be used to quickly distinguish people, vehicles, carts, animals, doors opening/closing, trees moving in the breeze, etc. Our system classifies objects into vehicles, individuals, and groups of people based on shape features (compactness and ellipse parameters), recurrent motion measurements, and speed and direction of motion. From a small set of training examples, we're able to classify objects in similar footage using a Fisher linear discriminate followed by temporal consistency.

Adding Cameras to the VMS

In the DVS environment, depending on the sophistication of the VMS software, you aren't always adding a camera, you're adding an IP address along with its ID and password as well. Dedicated VMS software suites for specific cameras may simplify the expandability process by creating a proprietary link between the VMS software and the add-on devices. The advanced VMS systems use the unique IP address in the IP camera and/or digital video encoder to add the device into the suite, assuming of course that there's a driver for that specific type of camera and/or digital video encoder.

■ ■ ■ ▬▬▬▬▬▬▬▬▬▬▬▬▬▬▬▬▬▬▬▬▬▬▬▬▬▬▬▬▬▬▬▬▬▬▬

TCP/IP Ports (again)

Whatever the network topology, it's crucial to the proper operation of any VMS software suite to have the assigned TCP/IP ports opened for accessibility of the client software, as within any organization most unused ports are closed to prevent unauthorized intrusion.

▬▬▬▬▬▬▬▬▬▬▬▬▬▬▬▬▬▬▬▬▬▬▬▬▬▬▬▬▬▬▬▬▬▬▬ ■ ■ ■

Whether adding an IP camera or digital video encoder, the IP address is how the device is added into the VMS interface (see Figure 7-7 for an example). If there's a digital video encoder that has multiple inputs, but is managed by a single IP address, then that digital video encoder requires a special driver for VMS to encode communications. It's best to check the VMS Web site for compatibility first and then, if supported, check drivers and special instructions.

User Management

The ability to create user accounts with clear hierarchical, multitiered permissions and privileges can be crucial, depending on the size of the implementation. For example, there may be security officers in three different buildings on a corporate campus. Those three security officers require access to the cameras within their building. In addition, there are three separate parking garages, each with select cameras, and those three parking garages have management offices that require access to their own parking garage cameras, but the security officers within each building also require access to their

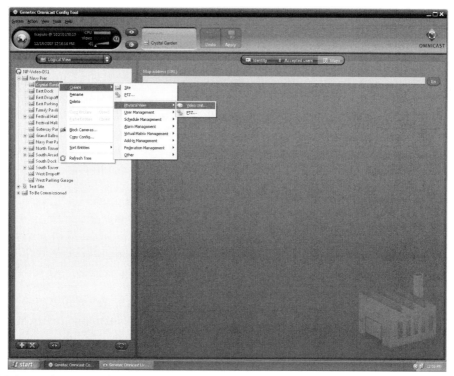

FIGURE 7-7 Drop-down menu for cameras/encoders.

individual parking garage cameras. At the top of the hierarchical chart is the chief security officer, who has an office on the top floor of Building 3.

Traditional CCTV would have been limited to each building, with monitors isolated for the security office and the parking garage cameras for the parking management office. Coaxial cable would have been run between each camera and a controller and/or video matrix at each office, and the chief security officer would be required to call each security office and/or parking garage management office for any real-time video surveillance or videotaped footage.

A DVS system, however, uses the IP network already in place and shares network resources between each building. All three buildings are connected though fiber and there's a single firewall that protects the entire IP network from the outside world. The corporate campus also shares a phone system, cable television, and access management systems.

The DVS conversion created three separate interim distribution facilities (IDFs) within each building that converged into the main distribution facility (MDF), which is also the server room, telecom room, and data center for the organization. The VMS server resides within the MDF, fully protected and environmentally conditioned, and the digital video encoders and new IP cameras are connected through the IP network.

However, providing everyone with access to all the cameras might not be the best idea for some facilities. Multitiered user management can segregate permissions and privileges not only by location but by functionality. The parking garage management office is provided full access to all their own garage cameras. The security office within that building and the chief security officer in his top floor office suite in Building 3 are also given access, but with no PTZ control. They can view but not change the area of coverage. There are also panic alarms installed throughout the parking garages for emergencies. Those panic alarms are integrated into the VMS system so in the event that the panic buttons are activated, the closest PTZ camera will automatically pan, tilt, and zoom (if the parking management changed the area of coverage for, say, a license plate) to that location within a second.

Depending on the VMS software, privileges can be as granular as locking the DVS viewer (so no one can minimize it and check their personal e-mail), locking the layout of the cameras being monitored, and camera presets to lock the area of coverage and the archived footage.

User Access

Typically, users are able to connect to the VMS system with a number of different access clients depending on authentication rights. The viewer client is usually the most feature-rich, depending on the privileges provided to the user through the administrative interface.

Users can usually pick and choose multiple camera views, varying designs, and maps of the select locations (as long as the maps have been designed and added into the system software), and create different camera presets for specific locations and objects of interest (e.g., egresses and ingresses, fields, intersections, etc.). Depending on the VMS software, they may also be able to create preset patterns and establish an automated panning of the select area of coverage.

All rights for users' access is run through the VMS server, centralizing the permissions for accessibility from anywhere, in the event that a user requires flexibility in the granular privileges provided or to be wiped from the system altogether. All professional VMS systems provide log files on access to the VMS server. These log files include user, time, date, and cameras that have been accessed. In Genetec, the privileges and permissions are controlled by the Config Tool; in Milestone, there's a Server Administrative login interface. Both of these controls require an administrative login.

As with many client/server applications, there are users and user groups. Individual user names can be assigned per individual accessing the VMS software, which is the best method of tracking. If there are large numbers of users in multiple locations, those individual users can then be assigned to select user groups to help with the administration and management.

There are typically four types of user groups to manage the video viewing and archives. Their general permissions include:

1. Administrators – Access for viewing live and archived video, PTZ control, and configuration

2. Live viewer users – Access for viewing live and archived video and PTZ control
3. Archive player users – Access is limited to archived footage only
4. Power users – Includes management privileges, but not overall administrative rights

Table 7-2 uses Genetec Omnicast as an example of the granular control provided for permissions and privileges for any newly assigned user or user group.

Archiving and Storage

DVS systems make it easier to use archived footage, which is primarily used for video forensics. Storage area networks (SAN), as discussed in Chapter 8, provide expandability and direct access to data, bypassing the server (except for security authentication into the VMS system). This speeds up the search for terabytes of raw video data on a multitude of hard drives.

There are several ways that video surveillance footage can be recorded, and those options are based entirely on the functionality of the VMS software. Economical software solutions, which also have a very limited device compatibility list, also offer limited resolution choices, continuous recording only, or may be linked to motion detection (if the device used supports it). The professional VMS suites offer these options as well as scheduling and user accessibility at a granular level, PTZ controls, numerous presets and patterns, select area motion detection, redundancy archiving, load balancing, failover options, and more of what you'd find in most high-end enterprise applications.

Remote Access

Professional VMS software suites offer remote access to the system, usually using a Web interface. This makes using the VMS suite outside the assigned workstations more accessible, as no client software may be needed to view cameras. Gaining outside access into the VMS system, through the Internet via a router or firewall, requires both an outside (public) IP address and a designated TCP/IP port to enter the closed network. The router or firewall must be configured so that requests sent to the outside IP address and designated port are forwarded to the inside (local) IP address and port of the server.

The ability to access the VMS on any computer, anywhere, at any time, over the Internet is useful for outside management. In addition to the abilities of the actual VMS, there are a few third-party technologies designed to provide access to internal workstations from remote locations, enabling users to run their familiar workstation from afar. Microsoft Windows includes Remote Desktop in its OS, allowing for anyone with the latest version of Windows to log in to the Windows Server from any remote location and work on the server desktop as if they were there sitting at the console. Other solutions include such software packages as PCanywhere by Symantec and Virtual Network Computing (VNC), which is free software that works with Windows, Linux, and Mac (www.realvnc.com).

Table 7-2 Granular Permission and Privileges Options for Genetec

Applications

Live Viewer	Allows the user to run the Live Viewer application without the privileges described under Live Viewer privileges, PTZ controls, and General privileges.
Archive Player	Allows the user to run the Archive Player application without the privileges described under General privileges.
Config Tool (Administrator Tool)	Allows the user to run the Config Tool application, without the Config Tool privileges.
Macro Editor	Allows the user to run the Macro Editor application.
Web Live Viewer	Allows the user to run the Web Live Viewer.
Web Archive Player	Allows the user to run the Web Archive Player.
Software Development Kit	Allows the user to run applications written with Omnicast SDK.
Pocket PC (Mobile)	Allows the user to run the Pocket PC application.
Media Gateway	Allows the user to establish connections with the Media Gateway.
Uncompressed Video Filter	Allows the user to establish connections with the Uncompressed Video Filter.
Federation Server	Allows the user to establish connections with a remote system's Federation Server.
Synergis	Allows the user to establish connections with Synergis access control system.

Config Tool Privileges (Administrator Tool)

Always view all entities	Allows the user to view all entity configurations. With this privilege, the user automatically gains access to all sites in the Logical view. See User Access.
Site configuration	Allows the user to change the configuration of existing sites and the Logical view hierarchy.
Creation and deletion	Allows the user to create and delete sites.
Unit configuration	Allows the user to change the configuration of units.
Firmware upgrade	Allows the user to upgrade the firmware of units.
Creation and deletion	Allows the user to create and delete units.
Camera configuration	Allows the user to change the configuration of video encoders, except the general settings, the recording and dynamic recording settings, and the motion detection settings.
Video quality settings	Allows the user to change the video quality settings.
Recording settings	Allows the user to change the recording settings.
Motion detection settings	Allows the user to change the motion detection settings.
Analog monitor configuration	Allows the user to change the configuration of analog monitors.
Audio configuration	Allows the user to change the configuration of audio devices.
Serial port configuration	Allows the user to change the configuration of serial ports.
Digital input configuration	Allows the user to change the configuration of digital inputs.
Output relay configuration	Allows the user to change the configuration of output relays.
PTZ configuration	Allows the user to change the configuration of PTZ motors.
Creation and deletion	Allows the user to create and delete PTZ motors.
Hardware matrix configuration	Allows the user to change the configuration of hardware matrices.
Creation and deletion	Allows the user to create and delete hardware matrices.
Schedule configuration	Allows the user to change the configuration of schedules.
Creation and deletion	Allows the user to create and delete schedules.

Table 7-2   Granular Permission and Privileges Options for Genetec—Cont'd

Config Tool Privileges (Administrator Tool)

Event and action configuration	Allows the user to change the configuration of custom events and actions.
Creation and deletion	Allows the user to create and delete custom events and actions.
Alarm configuration	Allows the user to change the configuration of alarm entities.
Creation and deletion	Allows the user to create and delete alarm entities.
Delete alarm instances	Allows the user to delete alarm instances before they're due to be deleted.
Macro configuration	Allows the user to change the configuration of macros.
Creation and deletion	Allows the user to create and delete macros.
Camera sequence configuration	Allows the user to change the configuration of camera sequences.
Creation and deletion	Allows the user to create and delete camera sequences.
CCTV keyboard configuration	Allows the user to change the configuration of CCTV keyboards.
Creation and deletion	Allows the user to create and delete CCTV keyboards.
Access control system configuration	Allows the user to change the configuration of access control systems.
Creation and deletion	Allows the user to create and delete access controls.
Monitor group configuration	Allows the user to change the configuration of monitor groups.
Creation and deletion	Allows the user to create and delete monitor groups.
Camera group configuration	Allows the user to change the configuration of camera groups.
Creation and deletion	Allows the user to create and delete camera groups.
Viewer layout configuration	Allows the user to rename the viewer layouts.
Deletion	Allows the user to delete viewer layouts.
Backup operator	Allows the user to perform backups operations.
Modify logical IDs	Allows the user to change the logical IDs of entities.
Plug-in configuration	Allows the user to change the configuration of plug-ins.
Creation and deletion	Allows the user to create and delete plug-ins.
View application connections	Allows the user to view which applications are connected.
View video connections	Allows the user to view all video connections.

Live Viewer Privileges

Change display	Allows the user to change the displayed elements in the Live Viewer's workspace.
Change tile content	Allows the user to change the displayed entities in tiles.
Change armed tile content	Allows the user to change the displayed entities in armed tiles.
Arm/Disarm tiles	Allows the user to arm and disarm tiles in the Viewing pane.
Change tile pattern	Allows the user to change the tile pattern of any viewer layout he has access to.
Change layout selection	Allows the user to change the selected viewer layout in each Viewing pane.
Edit/save layout configuration	Allows the user to edit and save the layout tabs in each Viewing pane.
Start/stop guard tour	Allows the user to start and stop the guard tour.
Edit guard tour dwell time	Allows the user to change the guard tour dwell time.
Acknowledge alarms	Allows the user to acknowledge alarms.

(Continued)

Table 7-2   Granular Permission and Privileges Options for Genetec—Cont'd

Live Viewer Privileges

Forward alarms	Allows the user to forward alarms and to set alarms autoforward.
Alarm snooze	Allows the user to make alarms snooze.
Audio (listen/talk)	Allows the user to use the audio controls.
Access digital zoom	Allows the user to use the digital zoom controls.
Do instant replay	Allows the user to use the instant replay controls.
Execute macros	Allows the user to execute the macros.
Change macro hot keys	Allows the user to change the macro hot key mappings.
Local recording	Allows the user to record locally on the PC hard disk.
Record manually	Allows the user to start and stop recording manually in the Live Viewer.

PTZ Controls

Perform basic operations	Allows the user to use the basic PTZ commands.
Change focus and iris settings	Allows the user to play with the focus and iris controls.
Use presets	Allows the user to use the camera presets.
Edit presets	Allows the user to change or rename the camera presets.
Use patterns	Allows the user to run the camera patterns.
Edit patterns	Allows the user to change or rename the camera patterns.
Use auxiliaries	Allows the user to use the auxiliary controls.
Edit auxiliaries	Allows the user to rename the auxiliaries.
Use specific commands	Allows the user to use the PTZ-specific commands and the menu mode.
Lock PTZ	Allows the user to lock the PTZ.
Override PTZ locks	Allows the user to override PTZ locks.
Add bookmarks	Allows the user to add bookmarks in Live Viewer and Archive Player.
Edit bookmarks	Allows the user to edit bookmarks.
Delete bookmarks	Allows the user to delete bookmarks.
View a camera on an analog monitor	Allows the user to connect a camera to an analog monitor.
Block camera	Allows the user to deny video connections to a camera from other users.
Send messages	Allows the user to do the "send a message" action.
Send sounds	Allows the user to do the "send an alert sound" action.
Send e-mails	Allows the user to do the "send an e-mail" action.
Send on serial ports	Allows the user to do the "send a string on a serial port" action.
Execute custom actions	Allows the user to execute custom actions.
Save and print snapshots	Allows the user to save or print snapshots.
Manually trigger an alarm	Allows the user to trigger alarms manually.
Start client application on a remote directory	Allows the user to view federated entities by connecting directly to the remote directory.
Control camera sequences	Allows the user to pause and step through the camera sequences.
Export video files	Allows the user to export video files.
Change own password	Allows the user to change his password.
Protect video from deletion	Allows the user to protect video from deletion.
Remove video protection	Allows the user to remove video protections.
Delete video files	Allows the user to delete video files.
Change application options	Allows the user to change the settings in the Options dialog.
Change client views	Allows the user to change the appearance settings of the application. Without this privilege, the user can't move the application window and can't log out.

Generally, these technologies fall into one of two categories: file or application access and access to the entire computer via a remote desktop. File or application access allows you to access your files or application from a remote location using the Internet. You can access a shared folder that contains the files or log in to the application directly using a remote client. Remote control of the desktop brings the entire desktop of the workstation onto the computer currently being used (see Figure 7-8). A remote desktop brings the entire working environment of the computer over the wire to the active workstation, eliminating the need for synchronizing files because you're essentially working on the remote computer.

Windows Server includes Terminal Server functionality, which allows for remote management of the system, using a remote desktop, for administrative purposes. If the feature is to be used for application sharing, there's an additional license required. However, VNC is a free application that provides the same functionality without limitations. Either way, both require open access to select TCP/IP ports to give remote access to internal resources (port 3389 for Windows Terminal Server/Remote Desktop and ports 5800 and 5900 for VNC).

VNC has a server and a viewer component. Its server allows other computers to connect remotely and the VNC viewer connects with the server from any client machine on which it's installed. VNC connections over the Internet aren't secure, and using an encrypted Virtual private network (VPN) connection is recommended.

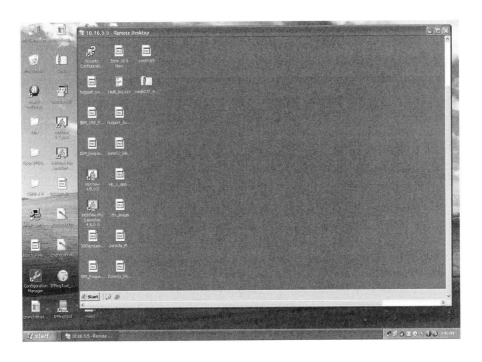

FIGURE 7-8 Remote Desktop, showing a Windows Server machine within a host computer.

Mobile

The ability to view video surveillance footage on a handheld device is ever growing in both options and quality (see Figure 7-9). Obviously, the biggest issue is the limited size of the mobile device screen, making it difficult to see much detail with more than one or two cameras visible. There's also the issue of functionality of cellular networks, both in bandwidth and quality, in any location mobile access is required. Mobile networks aren't all created equal and may require more than Internet access and VPN capabilities from the mobile device to gain access to a VMS server. The VMS must also be able to recognize the user accessing the server and provide the appropriate bandwidth solution for proper display (e.g., similar to the mobile or Web options for Web sites).

FIGURE 7-9 Mobiscope mobile video surveillance software on a Windows mobile device.

Typically, mobile video surveillance is limited to only a few frames per second, based on the limitation of the network and the processing power of the device, but it allows the user to monitor scenes from a different angle if, say, they're deep within the actual crowd under surveillance. Mobile solutions vary in both features and cost, with the professional systems now integrating mobile solutions into their software suites.

Troubleshooting

The VMS is a client/server architecture and follows many of the same rules as any other client/server application with the possible exception of more granular software permissions and privileges options.

As with any troubleshooting, as depicted in Figure 7-10, you must first confirm there's power throughout the topology. If there are two switches and a router between the workstation and the server, make sure everything is powered and running. A simple ping test from the workstation to the server can verify clear connectivity, and if that fails, then ping the router and/or firewall. If that fails, then check the network connection at the workstation and any switch in between.

A PATHPING can also provide a network path from the workstation to the server when there are any changes made to the topology, such as switching to another router and/or firewall, that haven't been properly configured for the VMS system.

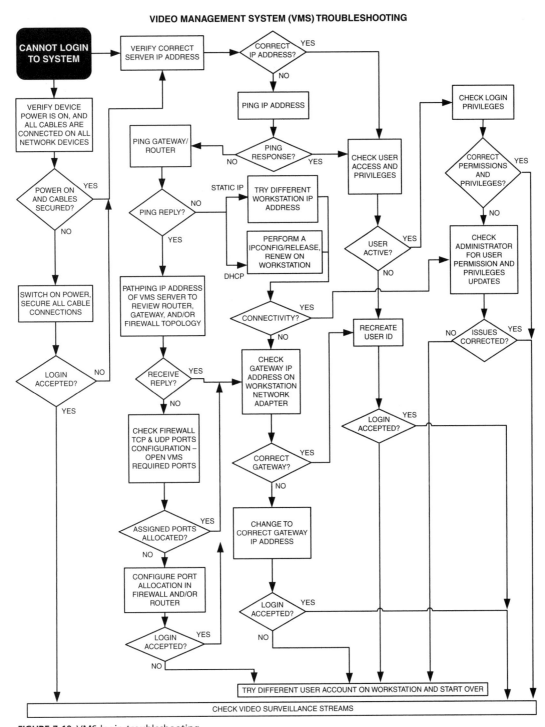

FIGURE 7-10 VMS login troubleshooting.

Depending on the complexity of the permissions and privileges, a quick diagnosis of authentication problems may be to just log in as another known-good account. If that's successful, then there may either be a mistakenly deleted account ID (it's happened) or the permissions were changed, making it impossible for that specific ID to log in to the system. Following the network troubleshooting suggestions in Chapter 4 will also assist in uncovering the problem.

Chapter Lessons

When choosing the right VMS solution, it's important to consider the following factors:

- Due diligence: Make sure the VMS system chosen can accomplish all the video requirements (e.g., compression technologies, integration, etc.).
- Research any compatibility issues with existing devices that require integration.
- Make sure there's enough storage capacity for all devices for the required time frame.
- Establish a secure, well-protected connection between all clients and the VMS server.
- Limit the number of users with access to the system from outside the working environment.

8 DVS Archiving and Storage

Introduction

Digital video security (DVS) archiving is the process of recording live video streams for later review. Archiving can be for a single day, 45 days, or more. It can be 24/7 or linked for motion detection to save storage space. Storage refers to the physical hard drive that holds the video recordings. The hard drive unit is usually called a digital video recorder (DVR).

DVRs come in all shapes and sizes, including the satellite and/or cable set-top box DVRs, the sole purpose of which is to record the streaming video provided for entertainment and education. There are video surveillance DVR systems that, like the television set-top box, provide the controls to review recordings on a digital or analog monitor directly off the internal hard drive. The larger implementations use a storage area network (SAN), which is a network specifically dedicated for the transporting of data for storage and retrieval. SAN architectures expand data storage on server hard drives by linking the storage capacity to other arrays through fiber optics.

A DVR is really just a computer with hard drives. The self-contained DVR has an operating system (OS) that's proprietary, simplistic, and inaccessible other than configuration capabilities (usually an open source version of Linux). These units may have limitations in capacity, not due to lack of space inside the unit as hard drives keep growing in storage capacity, but in what the embedded software can recognize. It's the same reason why a 2001 computer can't read a 1-TB hard drive.

Nevertheless, a DVR is still just the compilation of a central processing unit (CPU), video graphics card, memory, and hard drive space, which functions as a recorder of video surveillance footage for as long as there's space to hold the files and/or the schedule time. Long-term archiving and easy access to that archived video surveillance footage are the most powerful fundamentals of DVS.

DVS VMS Requirements

Typically, a DVS system includes a workstation where the video management system (VMS) is monitored and a VMS server where the software is installed and controls the cameras, video streaming, storage, and archiving. These two machines shouldn't be used for anything other than their primary purpose – to avoid performance degradation.

The primary reason to use two computers is that each handles different functions that continuously place demands on the machine. The computer workstation, which includes the VMS client software, requires exceptional video capabilities through a high-performance video graphics card. This video workstation displays multiple video streams for an extended period of time – whether during specific events, work hours, or 24/7 – and it demands horsepower designed to play video. The VMS server is built to handle recording video, which requires more computing horsepower and hard drive space. The server and workstation can only be one in the same with small implementations of two or four cameras, and even with that taxing the system it can't and shouldn't be used for anything other than a VMS server/workstation.

Power

There are two types of power requirements based on a number of factors, least of which is the size of the overall project. Power can refer to the amount of computing power, as in the number of microprocessors and their speed, or to the number of peak watts from the computing power supply. An enterprise-size implementation warrants enterprise equipment. The VMS server and archivers (additional hard drives) should be at the level of customer expectations. If the project includes 50 new cameras and the integration of 100 existing cameras, it requires quite a bit of horsepower to not only manage the 150 camera video streams and the number of users who are planning to log into the VMS to monitor the video, but also the consistent computing power needed to write video footage onto hard drives 24/7.

Storage Space

Hard drive space is inexpensive and easy to manage (compared to a VCR with magnetic VHS tapes), and today's DVS software offers the flexibility of how, what, when, and where to archive video footage using scheduling, motion detection, autotracking, and disk drive management. Although this chapter provides instructions for converting a computer into a DVR, keep in mind that there are many solutions available for any size implementation. The implementations in homeland security I've worked on required many terabytes of hard drive space and the enterprise level hardware to support it in the form of a SAN (see Redundant Array of Independent Disks).

The key to making a successful choice in archiving hardware is first analyzing the system requirements and using a storage calculator (such as the IPConfig Tool at www.jasc .com) to determine how much hard drive space is needed. A project with 100 cameras recording 24/7 at CIF resolution, using MPEG-4 compression, requires about 4 GB of space per camera, per day. That's about 400 GB of archived video footage per day, so a 1-TB hard drive would hold less than two and a half days' worth of footage. An implementation of 10 cameras would only require 40 GB per day, and thus that same 1-TB hard drive holds about 25 days of footage. However, if you choose to use megapixel cameras, which require more bandwidth, even using the new exceptional H.264 codec

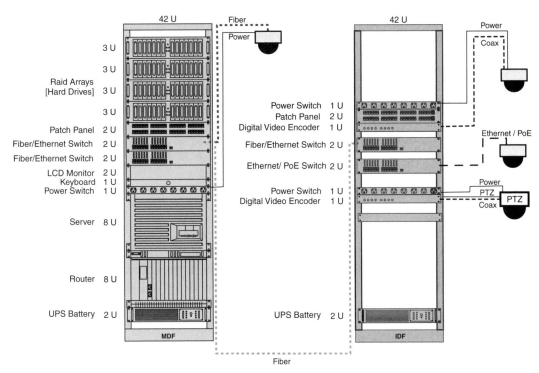

FIGURE 8-1 Rack-mounted VMS.

technology would increase the requirements considerably, depending on the megapixel size (see Chapter 2).

IBM, Hewlett-Packard (HP), and Dell are the industry leaders in providing enterprise level hardware, mostly in rack-mounted products (see Figure 8-1). The rack-mounted computing platform form factor provides a dramatic increase in the density of computing and networking devices by "stacking" those devices per unit of floor space.

The standardized rack-mounted form factor can be open – simply a few rails and a floor panel; it can be partially enclosed; or it can be fully enclosed within a cabinet under lock and key for additional security. A fully enclosed lockable rack, securely fastened to the floor and/or walls, severely restricts physical access by unauthorized personnel.

Standard rack-mounted computing implementations can be custom designed and built using various types and configurations and/or multiples of 1-U components. The standard 1-U unit is 19″ (482.6 mm) wide and 1.75″ (44.45 mm) tall. The most common rack-mounted computing form factor platform is based around a 42-U configuration, which means that a 42-U rack is capable of housing a maximum of 42 individual 1-U units.

There are many types of 1-U devices including network accessed storage (NAS), switches and routers, firewalls, power strips, uninterrupted power supplies (UPS), or even an LCD monitor and keyboard that neatly fold away. The multiple 1-U computers (servers and archivers) can be accessed using a keyboard, video, and monitor (KVM) switch, which allows access to each computer with the click of a button or a couple of keystrokes.

Additional rack-mounted options include lockable sliding rails, which allow a variety of equipment to slide in and out of sight without disconnecting it from the system. Rack-mounted equipment also provides easy access to the rear of the units, which may include inputs and outputs, power, and hardwired connectivity. Rack-mounted rails and cabinets also come with horizontal and vertical cable trays for protection and cable management, which makes it the best choice for large implementations where there may be dozens or even hundreds of cables to maintain.

Security

As mentioned in Chapter 1, the security of a VMS begins with the physical security of the equipment. The VMS server should be secured inside the data center or main distribution facility (MDF), behind lock and key, and in an environmentally controlled setting. The same holds true for any interim distribution facility (IDF), which is typically a network closet or a storage room converted into an interim network distribution facility.

The physical computer is just as vulnerable as the data inside the computer. What's the point of preventing anyone from gaining access to the data inside the vault if they can pick up the vault and leave? With tens or hundreds of thousands of dollars worth of hardware and software at risk, a rack-mounted cabinet provides the option to not only lock the cabinet door, but as with an exposed rack, to bolt the cabinet to the floor.

The Anatomy of a Computer

Whether the computer is a desktop PC sitting comfortably on a desk or a rack-mounted server in the data center or a DVR, they all have the following components in common:

1. Case enclosure
2. Power supply
3. Cooling fans
4. Motherboard
5. CPU
6. CPU heat sink and cooler fan
7. Video graphics card
8. Ethernet adapter(s)
9. Memory
10. Hard drives

There are two common ways computers are used: as a workstation and as a server. Workstations are desktops supplied to provide an individual with the power and speed

of all the computer has to offer from intranet to Internet to spreadsheets to word processing. A server is a computer that's usually behind the scenes, stored in a secure place.
It's the nucleus of the internetworking computer system. Smaller companies may have
one or two servers, while Fortune 500 companies may use thousands to handle their
ever-growing digital workload. These machines handle a plethora of applications and
systems designed to make business, the Internet, or a multiplayer network run better,
faster, and cheaper. As explained in Chapter 4, it's a hub in a wheel, connecting the
spokes to a tire, which makes it turn. It's a heart of knowledge, bringing cohesiveness
to the chaos of independent thinking. It's the one place that's always open and ready
to serve and protect you.

The only real difference between a workstation and a server is the OS and the software installed. The software on a server serves the many in the network, and because
of this kind of workload the Microsoft Windows network operating system (NOS) is
more robust with many more capabilities and is consequently more expensive.

The term server originates from its position in client/server architecture, where a
server is the program (or bundle of programs) that stands idle until called upon to fulfill
requests from client machines or nodes. In multi-tiered system architecture the server
may still rest on one machine while other components of the system such as databases,
authentication, and firewalls may fall onto other server machines for load balancing and
added security. There are database servers, file servers, Web servers, and application
servers, and, in the case of DVS, a video management system (VMS) software server.

The client, a typical workstation, runs any OS but can still access specific data and
resources from a centralized server through Ethernet. It may be running a different
OS altogether to handle unique or mission critical applications (e.g., Apple or Linux),
effectively meeting system requirements without affecting the client workstations.

Client/Server Architecture

Larger organizations already recognize the benefits of using a client/server system within
an enterprise-size environment. However, I believe it's especially important for smaller to
mid-sized companies. Microsoft's NOS is an affordable server OS (it comes in multi-tiered
pricing) that comes with a suite of solutions. This server can offer smaller companies the
same cost-efficiencies and improved processes as the goliaths of industry.

The Windows NOS includes extended built-in functionalities such as the ability to
plug and play more computers onto your network without having to touch it. This is
accomplished through dynamic host control protocol (DHCP), so when a colleague
visits from a satellite office out of town, you don't need to figure out how to get his
documents printed, e-mailed, or transferred – just plug him into the network and
the NOS will automatically assign him a temporary IP address to access network resources (see Chapter 3 for more information on DHCP). Windows NOS also includes
built-in administrative access from a remote location using Terminal Services and
Remote Desktop, which can also be used to view video surveillance footage remotely.

Upgrading Hardware for DVS

The cost of computers has plummeted over the past few years so upgrading a computer, beyond adding a faster CPU or video graphics card or additional memory and hard drives, may not be worth the effort. However, if the chosen computer meets the system requirements for a DVS workstation or server, with the exception of hard drive space or a faster video card, then it's definitely worth the effort, especially for smaller DVS implementations where thousands of dollars could be saved.

VMS Hardware System Requirements

Many manufacturers list the recommended minimum system requirements, and they truly are the very minimum. Computing power is cheap relative to a potential DVS implementation, so why use something minimalistic to represent the new system? The workstation is the most likely candidate for the final acceptance system testing plan, and since it's the station presenting the cameras of the DVS system, providing a work-horse for video can only benefit the overall system implementation.

Listing minimum system requirements is limited to the latest version of that specific VMS software, which can be upgraded every 18 months. To keep the current machine running for at least a few years, it's best to maximize the system requirements (see Tables 8-1 and 8-2).

Table 8-1   VMS Workstation System Requirements Table

OS:	Latest Windows Professional Version (32-bit)
Processor (CPU):	Dual Core @ 2.8 GHz or Higher (Intel or AMD)
Memory:	4 GB
Hard drive:	500 GB (for OS drive only)
Video memory:	DDR 1 GB + Dual Monitor
Power supply:	500 W + (1000 W + SLI Certified for dual PCIe video cards)
Sound card:	DirectX Compatible
DirectX:	Latest Compatible Version

Table 8-2   VMS Server Minimum System Requirements Table

OS:	Latest Windows Server Standard Version (64-bit)
Processor (CPU):	64-bit Quad Core @ 2.8 GHz or Higher (Intel or AMD)
Memory:	8 GB
Hard drive:	500 GB (for OS drive only)
Storage drives:	Varies
Video memory:	DDR 1 GB + Dual Monitor (Note: Nvidia didn't support Windows 2003 Server, so check for compatibility for procurement, or just buy ATI)
Power supply:	500 W + (1000 W + SLI Certified for dual PCIe video cards)
Sound card:	DirectX Compatible
DirectX:	Latest Compatible Version

Remember that the VMS workstation is taxed by *playing* video streams and the server by *recording* video streams, so the workstation needs high-end video graphics cards with the maximum memory possible, whereas the VMS server requires that maximum multiple processing power.

The Motherboard

The motherboard (also referred to as a main board or system board) contains the computer's core circuitry and components. On the typical motherboard, identification and configuration information is silk-screened on the planar surface. The embedded circuitry is on a chip or permanently printed or soldered onto the board as "traces" and most of them are made up of eight sandwiched layers or more.

Keep in mind that a computer is an interconnected system of individual, but compatible, components. The motherboard requires only a specific CPU form factor (and only up to a certain speed) and has a maximum memory capability.

The most common motherboard design for computers is the ATX, based on the original IBM AT motherboard. The components on all motherboards (or system boards) are

- A CPU or microprocessor
- Memory
- Basic input/output system (BIOS)
- Expansion slots
- Interconnecting circuitry
- Jumpers
- CMOS (BIOS) battery

The model number of the motherboard is typically located between the expansion slots on the bottom left-hand corner. Once the make and model are determined, check the manufacturer's Web site for the motherboard's specifications. This helps determine if upgrading is a possibility.

Motherboards have limitations. They can handle only so much memory, a CPU with only so much power and speed, and only so many hard drives.

■ ■ ■ ━━━

System Combos

If upgrading is the course of action, computer retailers such as newegg.com, tigerdirect.com, and microcenter.com offer specials on CPU and motherboard combos, which guarantee their compatibility. A VMS workstation would have similar specifications to a high-performance video gaming system. These computer retailers also offer bare-bones specials, which are typically a low-cost bundle of components minus the computer case and power supply. Upgrading computers isn't an easy task. It requires research, experience, and patience and any opportunity to make the effort easier is always a good idea.

━━━ ■ ■ ■

Central Processing Unit

Any time a computer is required to churn streaming video 24/7, either for viewing or recording, there's a desperate need for power, and not just computing power, but overall power.

The CPU is the brain of any computer, whether it's a desktop computer, a server, or a DVR. The power and/or speed of the CPU determine how many computations the entire system will be able to process and at what speed. This includes all the background services, power distribution, fail-safe mechanisms, etc. Typically, the faster the CPU and the more power, the better the performance. When purchasing a new computer or upgrading to a new CPU, make sure the motherboard is compatible and check the power consumption requirements to see if the power supply can provide adequate power for the upgrade. Of the individual components on the motherboard, the CPU uses the most power with the video graphics card in second place.

Video Graphics Card

The video graphics card resides on the motherboard in its own designated add-on slot. This slot provides a unique, faster link between the video graphics card, the CPU, and memory. There's usually only one slot of its kind on the motherboard, although there are systems that can handle multiple cards (driven by the video gaming market). On a computer (a workstation or a server for streaming video) a high-performance video graphics card is a requirement, not only on the server, but on any other workstations that access that server to view the DVS streams.

There are many choices in video graphics cards, including the highly recommended dual monitor ports. Two monitors provide double the viewing capabilities as well as the ability to minimize one monitor for incident reporting while sill monitoring cameras on the other monitor. Nvidia and ATI are the industry leaders in video graphics processors. To determine video graphics card limitations, the cards come measured in onboard memory (e.g., 512 MB, 1 GB, 2 GB, etc.). The more onboard memory the card has, the better the video performance. The most common video graphics card slot on motherboards today is the (Peripheral Component Interconnect Express [PCIe]), which is capable of transfer rates of up to 1 Gbps. The PCIe works within channels for full-duplex communications, rather than parallel buses of the preceding Peripheral Component Interconnect (PCI) slots.

Scalable Link Interface and Crossfire

There are select motherboards, video graphics cards, and power supplies that are considered as either scalable link interface (SLI) or Crossfire certified. SLI and Crossfire are the same concept from the two leading video card manufacturers, Nvidia and ATI, respectively. This technology allows the motherboard to handle more than one high-powered video graphics card for a maximum of four monitors per workstation (see Figure 8-2).

FIGURE 8-2 Navy Pier Command Center uses two dual monitor Nvidia video cards to link four 46″ LCD monitors per workstation *(photo courtesy of Tim Herlihy and Navy Pier).*

Multiple monitors are an excellent way to make even CIF resolution video images larger by showing fewer video tiles per monitor. Resolution, both in the size of the video and the video display, are discussed in Chapter 2.

■ ■ ■

Video Graphics Card Drivers

Before making a purchase of a video graphics card upgrade, make sure the OS supports that hardware. You must use the designated device driver for that specific video graphics card and for that specific OS to achieve maximum performance. If no drivers or the wrong drivers are installed, the video imagery appears pixilated. This can reflect on the implementation, not the specific computer used to display the DVS system.

■ ■ ■

Expansion Cards

Hard Drive Controller Cards

An integrated drive electronics (IDE) or serial AT attachment (SATA) controller card typically fits into an open PCI slot on the motherboard and includes the built-in firmware controller to run additional hard drives on the same motherboard. These cards are basically an extension of the motherboard, adding the additional ports or slots for more hard drives. The PCI slot runs at 33.33 MHz with a 32-bit bus width for a maximum 133-Mbps peak transfer rate (as opposed to the 1 Gbps+ capabilities with PCIe), so even with an SATA controller card, the transfer rate is limited to the peak transfer rate of the bus.

DVR Card

There are many DVR PCI expansion cards to choose from. They all have some type of software interface to allow you to view video streams. A DVR card is a digital video encoder in a computer expansion card form factor. It's essentially a video capture card, and is very similar to what would be used to watch television on a computer (Figure 8-3). As digital video encoders, all DVR cards require analog NTSC (or phase alternating line; PAL) video inputs.

DVR cards are typically designed to be used in the motherboard's PCI slot, which has a direct channel to the south bridge chipset. This PCI bus on any motherboard is a parallel bus; PCI slots can't transfer data at the same time and must take turns. In other words, the more PCI slots used on the motherboard, the better the chance of slowing down the overall performance of all the PCI expansion cards. It would be best for the performance of the video capturing and archiving to dedicate the PCI bus to a single DVR card, so an IDE or SATA controller card are great additions when more hard drives are needed. But using the motherboard's onboard hard

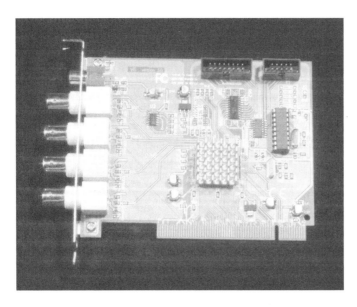

FIGURE 8-3 DVR card.

drive controller increases the DVR card's performance by dedicating the entire PCI parallel bus to the DVR Card.

DVR CARDS AND FPS

As discussed in Chapter 3, frames per second (fps) is the number of actual pictures captured within 1 second. DVR cards are limited in the number of frames per second based on the graphics encoder chipset used on the DVR card. This is one of the main reasons DVR card prices range from $50 to over $3000. The more economical DVR cards are rated based on fps and the number of video inputs. For example, if a DVR card has four video inputs and is rated at 30 fps, that doesn't necessarily mean that each video input will be digitized at 30 fps. Typically, that means 30/4 fps (or 4 divided by 30 fps). Each stream would have a maximum of 7.5 fps, so each stream becomes less fluid and "choppier." The smarter DVR cards determine how many video streams are being digitized and divide that number by the fps rating. DVR cards can take a single video input (or channel) or as many as 32 channels on a single card. So, if you use a 20-fps card and plug 16 cameras into the DVR card, each camera only plays and archives at 1.25 fps. If you use a 480-fps card with 16 cameras, each camera gets 30 fps and video playback is very smooth.

The more frames per second, the more powerful the card and the smoother the video. For best performance, select the DVR card that has the highest total fps to get a minimum of 15 fps per camera.

ADDING A DVR CARD TO A COMPUTER

Before adding a DVR card to a computer, make sure your computer meets the minimum system requirements for that particular DVR card. This includes making sure that there's a specific driver for the OS, that there's enough power to support the extra PCI card (expansion cards draw power through the PCI slot), there's enough hard drive space and/or the ability to add additional hard drives to store the recorded video from the added streams, and that you can reach the select analog cameras with coax cable from where the computer currently resides.

Before opening up any computer, make sure the power is off and that the power cable is removed. Many modern power supplies provide uninterrupted power to the motherboard to service the cooling fans inside the unit and even though it's a small amount and low voltage, it's best to disconnect all power before inserting (or removing) anything from the motherboard.

Adding a DVR card to a computer requires an open PCI slot, as depicted in Figure 8-4. There should be breathing space on both sides of the DVR card as it will heat up the system (make sure the system includes enough cooling fans).

FIGURE 8-4 Make sure there's an open PCI slot.

Once you've determined there's an open PCI slot, ground yourself before touching any electronic component by purchasing and wearing an electrostatic discharge protection wrist band (or just touch a metal object BEFORE touching any electronic component). Just a little static electricity is equivalent of being struck by lightning for any printed circuit board. Even if you don't completely fry the electronics, there may be a latency effect causing mysterious failures in the future.

Remove the blank PCI plate from the rear of the computer and then carefully insert the DVR card into the open PCI slot (see Figure 8-5). Confirm that it's secured in its PCI slot and locked onto the system, using the screw, and then power up the computer.

When Windows finally finishes booting, it will automatically recognize the addition of a new component. A ?, !, or x symbol appears over the device icon in Windows Device Manager, which can be accessed in a few ways:

1. Right-click on MY COMPUTER > PROPERTIES > HARDWARE > DEVICE MANAGER.
2. START > RUN and then type devmgmt.msc and press ENTER.
3. Open Computer Management in the Administrative Tools folder and it's listed as an option on the left pane.

The Device Manager provides any necessary information on each device on the computer. A hard drive, video graphics cards, network adapters – every component that the OS works with – are listed in this tree because each device needs a driver to interface between the embedded hardware firmware and the OS.

If an "x" appears, it means the device is disabled. Right-click on the icon and pick Enable. If the device can't be enabled, then the driver isn't functioning properly or may be incompatible. If an exclamation point appears, it signifies that a driver has been installed, but isn't functioning or may be the wrong version.

When the question mark appears it means that the OS recognizes there was a new component installed, but it doesn't know what it is (see Figure 8-6). You should insert

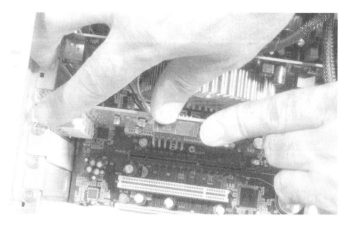

FIGURE 8-5 Insert DVR card into open PCI slot.

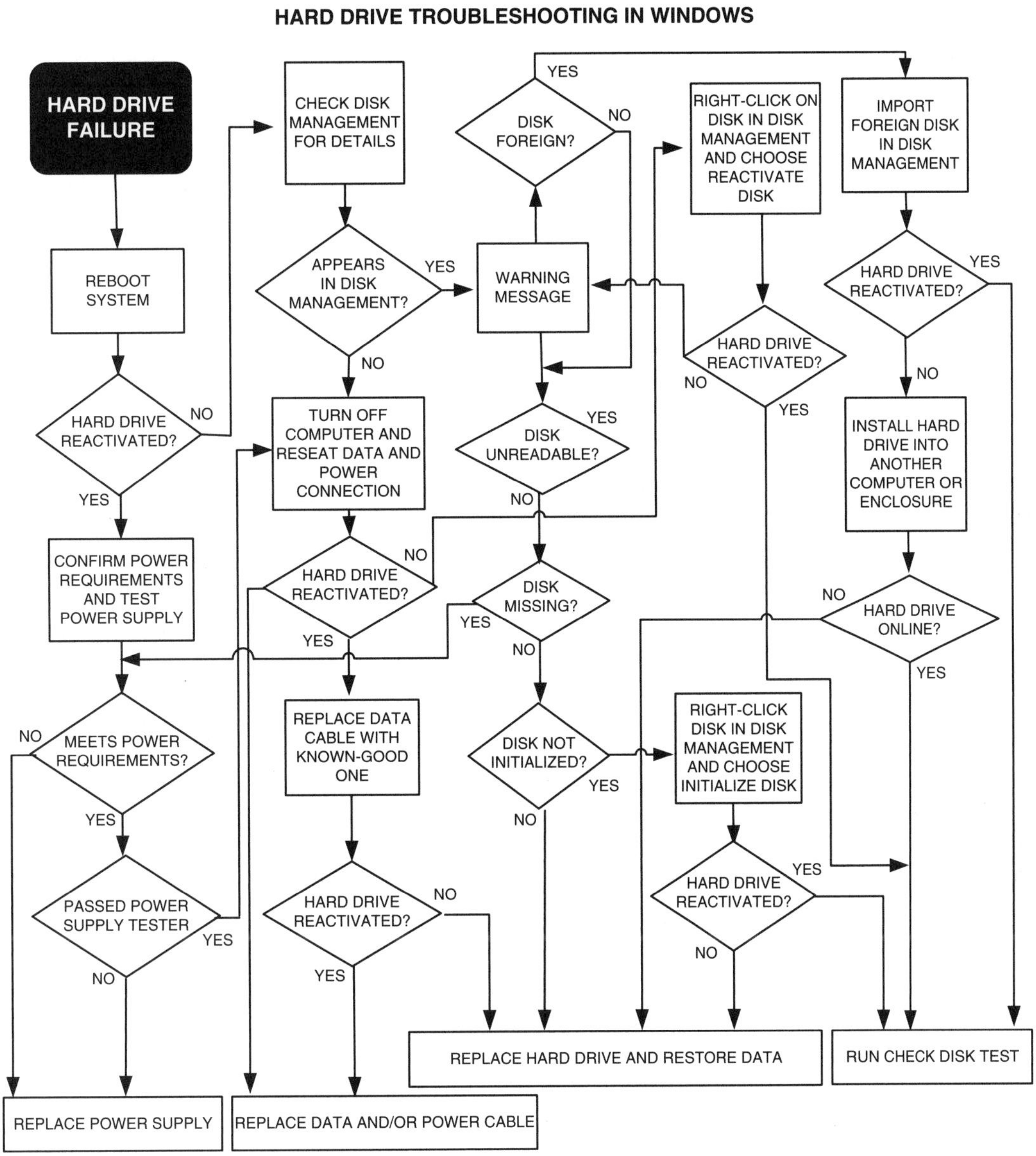

HARD DRIVE TROUBLESHOOTING IN WINDOWS

FIGURE 8-6 The Device Manager recognizes the insertion of a new device, but it doesn't know what it is.

the CD (with the drivers) or download the latest version of the drivers from the manufacturer's Web site and proceed with the installation of the drivers and software.

Upon completion of the installation (Figure 8-7), the question mark (or other symbol) should disappear from the Device Manager, indicating a successful addition into the system. Then it's time to set up the software with the cameras.

Keep in mind that this additional DVR card may not be recognized within a third-party VMS software application and vice versa. Typically, the software that comes with the DVR cards is dedicated to the firmware within the actual DVR card.

Power Supply

The typical computer power supply converts 110 VAC (or 220 VAC) into 12 VDC and 5 VDC to power the various components inside the system. The power supply includes a number of different connectors specifically designed for each component, so make sure that the chosen power supply (existing or an upgrade) can support all the components in the system and includes the correct amount

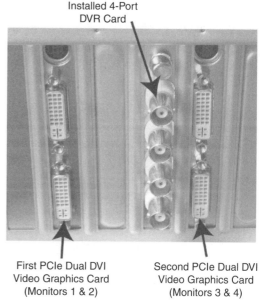

Installed 4-Port DVR Card

First PCIe Dual DVI Video Graphics Card (Monitors 1 & 2)

Second PCIe Dual DVI Video Graphics Card (Monitors 3 & 4)

FIGURE 8-7 Installed DVR card.

of connectors for all devices. Those connectors should include power cables for high-performance video graphics cards, SATA drives, and additional CPU power.

For example, if the system currently has two SATA hard drives and two more need to be added, make sure that the power supply supports those two additional drives and includes two additional SATA power cables. If the power supply is adequate for the two added hard drives, but there's only one additional SATA power connector, there are IDE to SATA power adapters that can convert an IDE Molex power connector for SATA power. In a SAN environment this becomes a question of multiple units and a UPS to provide enough power to the system to generate a soft shutdown, rather than a hard shutdown at the loss of power. Nothing will damage a hard drive or the contents on the hard drive like a hard shutdown or reboot.

Hard Drives

There are two types of common hard drive interfaces. The IDE standard interface is still used, but has given way to the faster, more energy-efficient SATA. The IDE interface is limited to a single master and slave per IDE port. An IDE drive may go up to 72,000 rpm when reading and writing to the hard disk, but the IDE interface has a maximum transfer rate of 133 Mbps. The SATA can reach over 10,000 rpm with a transfer rate of 3 Gbps and is recommended for both servers and workstations (Figure 8-8).

There are two ways to add another hard drive to your computer and they're based entirely on the computer's specifications. Larger motherboards may have multiple SATA

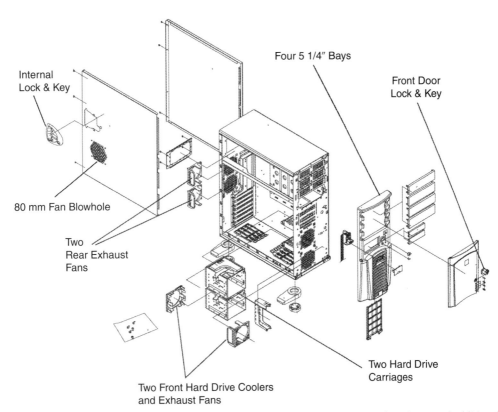

FIGURE 8-8 High-performance computer case with extra hard drive compartments and carriages and additional cooling fans.

ports (see Figure 8-9), providing the ability to add six or more hard drives to a single computer. If there's a single IDE port on a motherboard, that's typically used for a DVD drive as the master. A hard drive can be added as a slave if the need warrants additional space outside the SATA ports.

Figure 8-10 shows the SATA ports of my workstation with one open for a new hard drive. Most workstations have a single hard drive, so if it's a system that supports SATA hard drives, there's an excellent chance there are ports available for expansion without having to add an expansion card.

Adding an Additional Hard Drive

To convert an existing or new computer into a DVR, it should ideally have an additional hard drive for the archived video space to separate that function from the hard drive running the OS and VMS software.

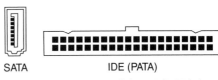

FIGURE 8-9 SATA port and the IDE (PATA) slot.

FIGURE 8-10 Close-up of the SATA ports on the motherboard.

Legacy computers running IDE hard drives are limited to two hard drives to the primary IDE connection on the motherboard (see Figure 8-11). If only the master position is used, there's room to add a slave hard drive. However, because it's an older motherboard it's best to first research the maximum hard drive size the motherboard's BIOS can recognize, with or without a BIOS upgrade (check the motherboard manufacturer's Web site). The recognizable hard drive size may only be limited to a specific size, so even if you install a hard drive with twice the size, if it reads it at all it may only see half of its full capacity. Again, for a small implementation of a handful of cameras that may be adequate space. So you need to be aware of the DVS requirements and match them accordingly.

More contemporary motherboards provide multiple SATA ports and in all likelihood, unless already used as a storage and/or backup system, there should be an open port for another dedicated archive hard drive. Check the motherboard manufacturer's Web site to determine the maximum recognized hard drive size and if there's a BIOS upgrade. Then decide how many and what size hard drives the system (including the power supply) can handle.

DVS requirements have a tendency to change, knowingly and unbeknownst to all parties, based on what appeared

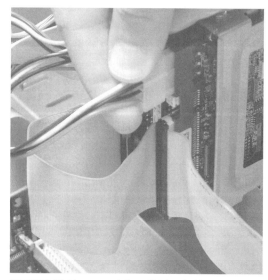

FIGURE 8-11 Master and slave IDE hard drives.

to be just a few minor changes. This is what makes change control so important (see Chapter 9, Project Implementation). Even a minor change, based on inaccurate information gathering and discovery, such as changing the bit rate of the video streams on a couple of cameras, can greatly affect the hardware requirements. This is why it's important to lock into those DVS requirements whenever possible and/or make sure there's a constant communication channel between all stakeholders to ensure there's quality control, customer satisfaction, and out-of-scope specifications (see Chapter 9).

As depicted in Figure 8-12, there are extra compartments for additional hard drives in all standard ATX computer cases and if not, a 3½" hard drive can be added into a couple of 5¼" DVD drive bays with a bay adapter (see Figure 8-13).

Ideally, video computers should have the largest and most reliable hard drives designed and proven for video storage and playback. The best system keeps the main hard drive that holds the OS and databases separate from the hard drives used to archive video footage. Although it's possible to have everything on a single hard drive, I personally don't recommend it for many reasons, least of all is if the hard drive with the OS is corrupted by a Trojan and/or virus, you'll lose all the video archives. If they're on a separate drive they will be far safer from attack. Stick with hard drives that have 7200 rpm or faster, which is common for most modern drives, but stay clear of 5400-rpm models – they may be less expensive, but they also provide less performance. For all larger DVS archiving and storage implementations, a redundant array of independent disks (RAID) is a must.

Redundant Array of Independent Disks

A RAID is a method of combining two or more hard drives in a format that enhances reliability and/or performance (see Figure 8-14). There are many types of RAIDs, but RAID 0 (zero) and RAID 5 arrays are more popular as they can "stripe" information

FIGURE 8-13 Adding a 3½" hard drive into a 5¼" DVD drive bay using an adapter.

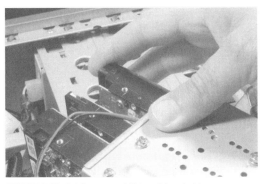

FIGURE 8-12 Adding an additional hard drive.

across multiple drives for maximum performance. When upgrading, consider purchasing a computer with built-in RAID functionality (usually listed as a feature of the motherboard), which allows you to use RAID formatting from at least a pair of identical hard drives. Although many systems use RAID 0 arrays, they're very risky. If one of the drives in a RAID 0 array fails, they all fail. See Table 8-3 for a list of available RAID levels.

Memory

Random access memory (RAM) also comes in a few flavors, but all of them appear to look the same. However, there are slight differences to consider when upgrading your computer because they

FIGURE 8-14 RAID array.

Table 8-3 Available RAID Levels

Level	Description	Minimum No. of Hard Drives
RAID 0	RAID 0 offers improved performance through *striping* data over multiple hard drives, and by doing this, increases the storage capacity, but there's no redundancy or fault tolerance. When a single hard drive fails, it destroys the entire array.	2
RAID 1	By adding a *mirroring* function, RAID 1 provides fault tolerance from disk failure of all but one of the drives in the array. There's also an increase in read performance when using a multithreaded OS (seeking two places at the same time).	2
RAID 2	Using what is called *hamming code parity,* the disks in the array are synchronized and striped in single bytes and/or words.	3
RAID 3	RAID 3 offers fault tolerance similar to RAID 5, but because the byte-level stripe is smaller than a file system block, the read and write to the array act as a single drive. All the drives must have synchronized rotation for this to work properly, but if one hard drive fails the performance doesn't change.	3
RAID 4	Similar to RAID 3, but uses block-level striping instead of byte-level striping. Data files can be distributed across multiple disks, so as each disk operates independently. This allows for parallel I/O requests reducing data transfer speeds but offering error detection through dedicated parity on a separate, single hard drive.	3
RAID 5	The major benefit of RAID 5 is that the distributed parity requires all drives but one to be present to operate, so if there's a single hard drive failure the array isn't destroyed. Using a distributed parity, upon a hard drive failure, reads are	3

Table 8-3 Available RAID Levels—Cont'd

Level	Description	Minimum No. of Hard Drives
	recalculated and the end user is unaware of a hard drive loss. However, if a second drive within the array fails there can be data loss and the whole array is vulnerable and will have reduced performance until the hard drive is replaced and the data on the failed drive are restored.	
RAID 6	Similar to RAID 5, but fault tolerance includes the ability to function even with two failed hard drives in a single array. Similar to RAID 5, the whole array is vulnerable and has reduced performance until the hard drives are replaced and the data on the failed drive are restored. The larger the hard drive capacity, the longer the restoration of the hard drives takes. Dual parity adds extra time to the array restoration without the data being at risk if another hard drive fails before the complete restoration of the previous failed hard drives.	4

have different form factors and speeds. Whether using the computer for a DVS server or workstation, it's best to max out the memory on the motherboard, but keep in mind that the standard 32-bit Windows OS can only read up to 4 GB of RAM. If the software or hardware requirements call for anything more than 4 GB of memory, it's recommended to upgrade to a 64-bit system.

The Network Operating System

The NOS is the brain of the server; without it, it's just a computer. It offers applications and utilities that do business faster and better. There are only a few popular choices – Novell, UNIX, Linux, and Windows. The complexity of NOS forces a simple overview of the features and benefits.

Although Linux offers a free download and a plethora of available features to function as a NOS, the most attractive aspect of this Microsoft product is the monolithic support structure and community. All VMS software works on Microsoft OS. Microsoft takes sophisticated tools and applications and makes them usable and affordable, so that any size company can take advantage of a complete suite of server utilities and applications. However, many people have a bias and prefer other systems to Microsoft or simply don't like Microsoft. Years ago, one of my clients chose a Web servicing company that slapped together an assortment of custom and obscure technologies from which they created four unique Web sites. Providing them with a shared Oracle database gave my client a more cost-effective database solution, but negated other features (for security reasons they were told) and locked them into using this particular Web services company. My client was unconcerned about this arrangement until the company found itself tied into multiple approvals and design changes to their Web site at $200 an hour. Arrangements like this can become very expensive. The IT director hated Microsoft and their products with a passion (there's one in every crowd). This prejudice steered the company to technologies that had no immediate support structure set into place. At one point they attempted to internally convert the original NOS, Application Server, and dynamically generated Web sites from one non-Microsoft platform (Linux/Jrun/Java) to another non-Microsoft platform (Novell/Websphere/Java), but they had a problem finding a consultant with the appropriate skill set who was immediately available.

After they spent a month of searching, experiments, and multiple dead ends, I walked into their building armed with a developer copy of the Windows Server NOS and asked for one server machine with a static IP address and access to the Internet. That was at about 1 p.m. and by 5:30 p.m. (after some BIOS upgrading), I had installed and configured the Windows Server NOS, configured a Web Server and Application Server, and installed all four Web sites. The director was bewildered at the rapid deployment, so much so that he asked me to show him the process step by step.

Over 95% of computer users utilize a Microsoft OS, which also means that most of the VMS software (as most all software) is developed for the Microsoft platform. It's an OS that's understood and has become more intuitive and an intricate part of our daily lives; thus it's the better choice for small or mid-sized companies with limited time and resources (human and otherwise).

Typically, a networking environment opens up shared resources such as files, printers, and an Internet connection. The Windows Server NOS provides a configuration wizard that gets the server up and running within minutes. This also includes a few functions that are required for remote viewing of the VMS software such as a Web server or Terminal Services.

NOS provides more features than will ever be used, but there are a few select features beneficial for a DVS deployment.

Administrative Tools

Microsoft Windows, both for workstations or for servers, comes with a set of computer management tools that help you manage the machine. These are very important and should be pinned to the Start Menu. That can be done by navigating to the Administrative Tools folder in the Control Panel:

START > CONTROL PANEL > ADMINISTRATIVE TOOLS

Inside the Administrative Tools folder is an icon named Computer Management. Right-click on that icon and choose Pin to Start Menu.

The Computer Management console is a pre-configured interface with a number of administrative and troubleshooting components to save time. This console can also be accessed by doing the following:

START > RUN and Type COMPMGMT.MSC

Scalability

One of the more important aspects of choosing a server over a workstation to function as a DVR is that the server and its NOS are more scalable. Eventually, others may wish to monitor the video surveillance streams or access archives and that can add extra burden to the single workstation. A server makes it easier to have a centralized location for recording and security procedures, including authentication, accessibility, and control. As the demand for VMS access increases in the client/server environment, the installation of the client software onto another workstation is all that's needed.

IP Cameras

IP cameras don't require a DVR card, because they're already generating digital video through the built-in encoder. When choosing an IP camera system, it's best to focus all the research efforts on the VMS software. The VMS must be able to support all the chosen IP camera makes and models to integrate with their features and store and manage the video surveillance archives on the local hard drives. Even with a DVS solution including all IP cameras, there's still a computer somewhere that becomes a DVR; it's just software driven.

Network Accessibility

Control of the DVS network begins with authorization and authentication by limiting the number of people who have access into the DVS system through ID and passwords. In addition to controlling log in, granting access to select physical media access control (MAC) addresses only allows unique computers to join the network. This can be cumbersome in larger implementations but is important to protect the security assets and bandwidth.

Disabling the IGMP protocol for anyone except those granted by MAC address will also stop anyone from pinging the network and creating a Denial-of-Service attack. Also, as I've mentioned in Chapter 4, don't use DHCP on a DVS network as this provides a dynamic IP address to devices (even laptops), giving them all a way to join the subnet and gain access into the network or at least the IP scheme and subnet.

Firewall

A computer firewall is similar to a physical firewall, which keeps fire from spreading from one area to another. In a computer network, it's designed to block unauthorized access while permitting authorized communications. It's hardware and/or software configured to permit, deny, encrypt, decrypt, or proxy all (in and out) network traffic between different security systems based upon a set of rules and other criteria.

Firewalls, when properly configured, examine data that pass through and block the data that don't meet specified security criteria. Firewalls use several different methods of access control like packet filtering, where it examines each packet entering or leaving the network and takes action based on user-defined rules. As an application gateway, the firewall assigns security mechanisms to specific applications such as remote access applications like HTTP or FTP servers. The firewall can also provide a circuit-level gateway by applying security mechanisms to TCP or UDP connections once established. As a proxy server, the firewall intercepts all messages entering and leaving the network, effectively hiding the true network addresses.

There's no need to be concerned with hackers with a properly configured firewall. However, there are many ways to gain access to the system without having to intrude in real time by using Trojans, viruses, and worms.

Malicious Software

A Trojan is a file that has a hidden content with malicious intent. Trojans are typically encapsulated as something harmless and enticing, such as a game, video, or a picture, but once you execute this file (run it), the worm or virus is released onto the system.

Viruses are computer programs with the sole purpose of destroying data on computers. The virus may destroy what appear to be unimportant files until you attempt to use one of the programs or another feature of Windows, or it may erase all of your document files or corrupt the master boot record or complete registry file. Viruses are spread through executable files (.exe) downloaded from the Internet or installed through a flash drive. A virus can be disguised under the cloak of a Trojan, which is a carrier for the virus.

Worms replicate themselves, reaching over networks to multiple computers unprotected by firewalls. Worms come through e-mail, through Trojans, and even scripting code from visiting unsavory Web sites.

DVS Remote Viewing

Many VMS applications provide a means to view the client application remotely, either by a select configuration of a firewall or through a Lite Viewer, using a Web browser to simulate the VMS interface. This usually involves setting up a Web server to service Web pages by using a scripting language and an application server to gain access to the VMS software and archives. As a standard Web server, this only limits the client device to whatever can support the scripting application (e.g., Java, .NET), which also includes smart phones. There are also a number of select shareware applications or the free edition of RealVNC (www.realvnc.com/vnc/features.html), which offers a remote desktop type of environment for viewing an office computer from home.

Terminal Services and Remote Desktop

Terminal Services, which is part of the Windows NOS suite, provides the ability to host multiple simultaneous client sessions on Microsoft Windows Server NOS. Terminal Server is capable of directly hosting compatible multi-user client desktops also running on a variety of device hardware (e.g., Windows Mobile Professional smart phones include Remote Desktop to access Terminal Servers). Standard Windows-based applications don't need modification to run on the Terminal Server and all standard Windows Server 2003-based management infrastructure and technologies can be used to manage the client desktops.

Hard Drive Preventative Maintenance

DeFrag

It's important to understand that hard drives require preventative maintenance, not only to extend the life of the disk, but to also protect the data. The maintenance step missed most often is disk defragmentation. Also available within the computer management console, "defraging" collects data that have been peppered all over the disk into more compact locations for stability. This is important for hard drives that are continuously overwritten or if multiple files are deleted, as in a hard drive recording DVS video streams. It's best to defrag all the hard drives on a monthly basis.

Chkdsk

If there's any sign of hard drive problems, such as reduced performance or periodically not booting online, then run the Windows Check Disk (Chkdsk.exe) command line tool that checks volumes for problems and also attempts to repair those discovered problems. Chkdsk can repair problems related to bad sectors, lost clusters, cross-linked files, and directory errors. Chkdsk will only work if logged on as an Administrator or as a member of the Administrators group.

There are two ways to access Chkdsk:

(1) START > RUN and type CHKDSK.EXE and press ENTER.
(2) Right-click on the hard drive icon (in My Computer), chose
PROPERTIES > TOOLS > CHECK NOW.

Cooling Fans

Many people don't realize how warm a computer gets with all the components inside generating heat. One of my computers, which I'll confess doesn't have enough space for cooling fans, had one of the chipsets running at 159°F. Although computer electronic components are designed to withstand some heat for extended periods of time, experience has taught me that the more components you stuff into a computer case, the more fans it needs to extend the life of those components. Since there were no additional spaces for fans, I added a hard drive cooler for the larger hard drive (see Figure 8-15) and dual rear exhaust fans that dropped the temperature down to 135°F.

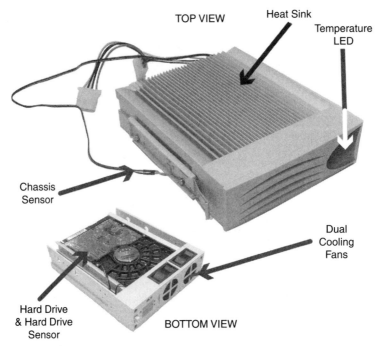

FIGURE 8-15 Hard drive cooler.

Troubleshooting

Troubleshooting the DVR Card

The process of troubleshooting a DVR card (Figure 8-16) is similar to any PCI expansion card. As with any new computer hardware, a driver is required. A driver is the software interface between the built-in firmware embedded within the device and the computer's OS. They're specific to each OS. In other words, if the specifications don't include a NOS, then it isn't supported, creating problems that can only be corrected with the proper device drivers or with a new device that's compatible. Many times this is done in the background as Windows may have the driver already built into the OS. If not, the device usually includes a CD-ROM with the appropriate drivers for the selected OS, and the proper VMS software to manage the new DVR card and the video streams. Even if the device includes a CD-ROM with the designated drivers per OS, it's always best to check the manufacturer's Web site for the latest device drivers.

There are a few common problems with DVR card installations, although each is very different, depending on the requirement. Assuming that the operating systems (both on the workstation and the server) support the DVR card, then and only then can troubleshooting begin with the simple verification of power. If there's no video present, make sure that the cameras are powered, the cables are secured, and the DVR card firmly inserted into the PCI slot. The PCI slot and the motherboard power the DVR card.

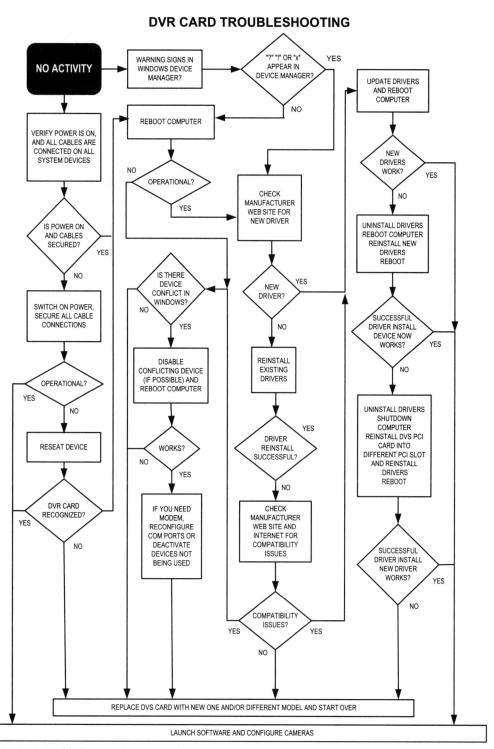

FIGURE 8-16 Troubleshooting the DVR card.

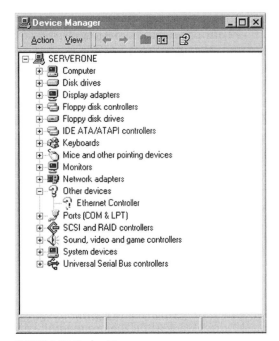

FIGURE 8-17 Device Manager.

Once all power and cables are verified and secured, the next step is to check and see if the ?, !, or x appears over the device icon in Windows Device Manager. There are a few ways to access the Device Manager:

1. Right-click on MY COMPUTER > PROPERTIES > HARDWARE > DEVICE MANAGER.
2. START > RUN and then type discmgmt.msc.
3. Open Computer Management and it's listed as an option in the left pane.

The Device Manager (Figure 8-17) provides necessary information on each device on the computer. Hard drive, video graphics cards, network adapters – everything that the OS interfaces with is listed in this tree.

If the DVR card tree is opened and the icon is covered with a ?, !, or x, the first step is to reboot. Rebooting can accomplish one of two things. It can reinstall the drivers once the OS boots up and, even though it may not require a reboot, sometimes this information is left out of the installation instructions.

Once the operating system is up and running again, if the ?, !, or x still appears over the DVR card icon in the Device Manager, then the next step would be to check the manufacturer's Web site for a new driver update or patch, depending on the OS. There's also a possible device conflict, where the DVR card is attempting to share resources with another device (such as a network adapter) and that other device doesn't like to share. This can be determined via the System Information Manager:

START > ACCESSORIES > SYSTEM INFORMATION > PROBLEMS/CONFLICTS >
START > RUN and then type discmgmt.msc.

If there's a conflict, the only option is to disable the conflicting device, only to confirm that the DVR card will become operational once the other device is disabled. Of course, if the other device is a network adapter or USB controller, there may be a problem. Unless the conflicting device is something absolute (e.g., video graphics card or hard drive), disable it temporarily to verify operation. If this brings the DVR card to life, then follow the detailed instructions on how to change computer resources for select devices in the computer BIOS or in Windows, depending where the conflict occurs. Of course, I highly recommend having someone with extension experience do this as you could cause serious damage.

Another trick involves an uninstall and reinstall. Make sure to close and disable all applications including the antivirus software, firewall, and even System Restore. All in all, the installation of a PCI expansion card is a mature process that typically doesn't require many hours of troubleshooting. If these steps were followed and there are still issues with the DVR card, then it's best to find another device that's clearly more compatible. It's very important to check online for reviews and forums for potential problems. A few years ago, I wrote a how-to book on building a server. In the process of building the system, I was having issues with the select video graphics card I chose. This Nvidia video card (industry leader) wouldn't work on an Elitegroup motherboard. I tried two other cards with no issues and troubleshooting became very frustrating. I Googled my problem and after about a half hour of research I discovered this pairing had a strange anomaly that required pushing down onto the video card as the computer was booting. As ridiculous as it sounded, I was desperate, so I gently pushed down onto the video graphics card and hit the power button. Sure enough, the video card came to life and was recognized thereafter. We're talking about different devices, designed and manufactured independently, and briefly tested under select configurations (you can't test every device on everything). The Internet is your troubleshooting friend.

Troubleshooting a Hard Drive

There's very little that can be done when a hard drive fails, but the key is finding out how failed it actually is. The simplest way to determine if a hard drive can be salvaged is to feel it as it boots up. If there's vibration, the internal disc is rotating. This means there's a chance to retrieve the data without the need of a professional (Figure 8-18).

The following troubleshooting recommendations are for a regular NTFS-formatted hard drive and not for a hard drive that's part of a RAID array. Don't use FAT32 formatting for DVS, as the system isn't as robust as NTFS and the files are limited to 2 GB in size. The RAID hard drive would need to be replaced (with the exact same hard drive) and the RAID software will recover the data (except for RAID 0). This doesn't pertain to the boot disk, only to additional hard drives used for archiving video.

There are two types of disks in a Windows environment. A basic disk is a physical hard drive that contains primary partitions, extended partitions, or logical drives. A dynamic disk includes the ability (but not a necessity) to create volumes that span multiple disks and create mirrored and RAID 5 fault-tolerant volumes. Single dynamic disks are also easier to move from one computer to another.

■ ■ ■ ▬▬▬▬▬▬▬▬▬▬▬▬▬▬▬▬▬▬▬▬▬▬▬▬▬▬▬▬▬▬▬▬▬▬▬▬▬▬

Whenever troubleshooting electronic components, always ground yourself from electrostatic discharges before handling anything and always turn off the power before disconnecting or connecting any device.

▬▬▬▬▬▬▬▬▬▬▬▬▬▬▬▬▬▬▬▬▬▬▬▬▬▬▬▬▬▬▬▬▬▬▬▬▬▬ ■ ■ ■

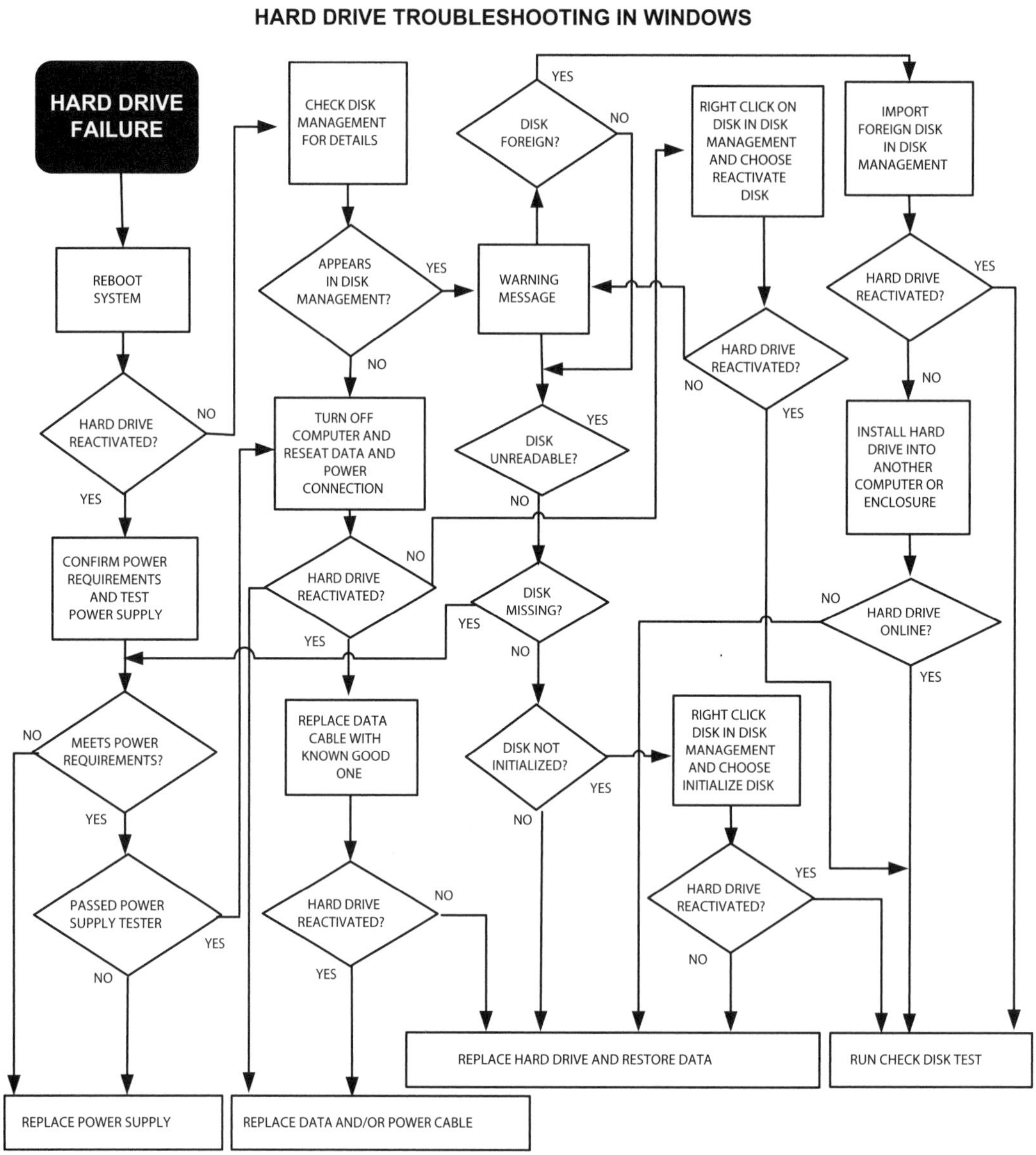

HARD DRIVE TROUBLESHOOTING IN WINDOWS

FIGURE 8-18 Hard drive troubleshooting.

There are a few ways a hard drive indicates it's having issues. The first is simply not appearing. The first step to determine the severity of the hard drive failure is a system reboot. If the hard drive returns in My Computer, then you must run a few select tests, one of which is testing the power supply for compatibility and/or failure. This is especially important as a defective power supply can damage components inside the computer, including the motherboard, hard drives, and memory. A power supply tester is an inexpensive addition to the troubleshooting arsenal and can give you instant results.

■ ■ ■ ━━━

Data Disaster Recovery

As long as the hard drive powers up and continues to spin, there are a few possibilities for bringing the hard drive to life, at least long enough to retrieve the stored data. The following methods have worked for me at various times to revive hard drives to retrieve the data inside.

1. Installing the hard drive into another computer
2. Adding the hard drive into an external hard drive enclosure
3. Carefully hitting the side of the hard drive with a clean flat surface
4. Place the hard drive in the freezer for an hour

━━ ■ ■ ■

If none of these methods works and the stored data are mission critical, then data recovery software, such as Stellar Data Recovery (www.stellarinfo.com), has worked wonders with a free trial version to determine if the data is salvageable.

Upon researching the computer's components, you may find that the power supply is below the recommended rating for the current system. At this point, the best course of action is to upgrade the power supply and then start the troubleshooting over again. Troubleshooting hard drives in Windows includes the use of the Disk Management tool, which is part of the Computer Management suite (see Figure 8.19). Again, to access Computer Management, go to START > RUN and type compmgmt.msc and then press ENTER.

The Disk Management console is listed in the left pane under Storage. Click on the Disk Management icon and the computer's storage devices will appear in the right pane of the console. The hard drive may appear in the Disk Management console with error messages that may help diagnose the problem. If the hard drive doesn't appear at all in the console, check to see that the hard drive's data and power connection is securely in place and then reboot.

The messages and/or status errors that appear in the Disk Management console include, but aren't limited to, the following:

* Foreign disk
* Disk unreadable
* Disk missing
* Disk not initialized
* Disk offline

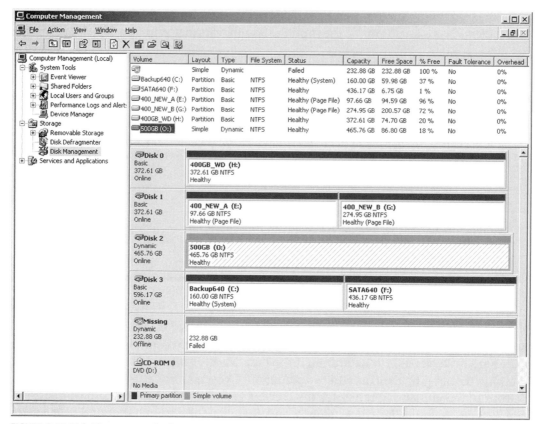

FIGURE 8-19 Disk Management in the computer management console.

A warning icon appears on disks that display the Foreign status, which signifies a moved dynamic disk, either to another port or from another computer. Dynamic disks aren't supported on Windows XP Home Edition or on portable computers.

To access data on the disk, you must add the disk to your computer's system configuration by importing the foreign disk (right-click the disk and then click Import Foreign Disks). An error icon appears and the hard drive is labeled as Unreadable when the disk isn't accessible. The hard drive may have experienced hardware failure, corruption, or I/O errors. Sometimes the unreadable disk failed and isn't recoverable, but for a dynamic disk this usually means corruption or I/O errors on the part of the disk, not a complete failure of the entire disk. In the Disk Management console, click on Action and then choose Rescan Disks and/or restart the computer to see if the hard drive's status changes.

The Missing status indicates a corrupted, turned off, or disconnected dynamic disk. Instead of appearing in the status column, the Missing status is displayed as the disk name. Make sure the hard drive is connected and powered and then open Disk Management, right-click the missing disk, and then click Reactivate Disk.

When the Not Initialized status occurs, it means that the hard drive doesn't contain a valid signature in the master boot record (MBR). Disk Management provides a wizard when a hard drive is first installed into the system that once followed, and will add this signature into the boot record. If the wizard was canceled before the hard drive signature was written into the boot record, then the disk status remains Not Initialized. Right-click the hard drive in Disk Management and then click Initialize Disk. The hard drive then changes to Healthy status.

Chapter Lessons

- Keep all computers, whether a VMS Server or a VMS workstation, in a cool, dry, environmentally controlled space.
- Recording and/or viewing video streaming requires maximum computing power. VMS servers should include more multiprocessing power, and the VMS workstations need high-performance video graphics cards to display the best quality video.
- When calculating storage space requirements, remember there's a difference between a bit and a byte or use a calculating tool such as IP Design Config Tool at www.jasc.com.
- Security of the VMS system begins with the physical security of the hardware. Keep all equipment under lock and key.
- Always ground yourself before handling any electronic printed circuit boards.
- Whenever implementing a large DVS system, always use a RAID array for video archiving to provide for fault tolerances.
- Computer systems can never have enough cooling fans to extend the equipment's life expectancy and prevent latency issues.
- When converting a computer into a DVR using a DVR card, always confirm the software and driver compatibility with the chosen OS.
- When hard drive problems occur, try installing the hard drive into another computer to see if it's recognized and mounted. Add the hard drive into an external hard drive enclosure and carefully hit the side of the hard drive onto a clean flat surface, or place the hard drive in the freezer for an hour and then try again.

9

Project Implementation

Introduction

A project is defined as a temporary endeavor with a beginning and an end that creates a unique product, service, or result. A project is also considered progressive, documented, and managed. Larger projects can be divided into manageable phases or subprojects, which make it possible to assign these parts to other individuals, departments within the organization, or subcontractors who may have better expertise on the subject matter and thus greatly reduce the risk.

This chapter focuses primarily on the management of a project and the specific deliverables required from previous digital video security (DVS) projects. The complexity of any DVS system warrants a structured foundation, no matter the size. Portions of this chapter may be considered overkill for small projects, but the information is here to enlighten any project manager on the proper steps and actions for certain situations should the need arise.

Project Management Institute and the Real World

Project management is the application of knowledge, skills, tools, and techniques to manage activities to meet project requirements. The Project Management Institute (PMI) (www.pmi.org) has developed a highly respected certification process for project management professionals (PMP) that suggests any project is accomplished through five stages or processes:

- Initiating
- Planning
- Executing
- Monitoring and controlling
- Closing

Initiating is the first step in the project management process and includes all of the steps and work necessary to create a project charter. A project charter is a document that provides the high-level details of the project including project title and description, the assigned project manager and his authority level, the project's goals and objectives, and the business case benefits. Also included is the senior management involved in decision making and the appropriate signatures for approval. Nothing should move forward without a written agreement signed by both parties, detailing a clear understanding of the project's goals and costs. In the PMI world, that document is called a project charter; out in the real world it may have different names and comes in different forms, but the need for a signature of approval remains the same.

The initiating phase (see Figure 9-1) may also include a conceptual design to be able to evaluate high-level costs, assumptions, and constraints, but is primarily about authorizing the project or phase.

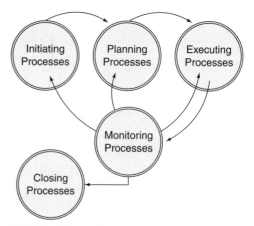

FIGURE 9-1 The various processes during a project life cycle.

■ ■ ■ ━━━━━━━━━━━━━━━━━━━━━━━━━━━━━━━━━━

PMO

A relatively new concept in the management of projects, the Project Management Office (PMO), is typically a formal structure that supports project management within an organization. It furnishes the policies, methodologies, and templates for managing internal projects or lends support and guidance to others in the organization on how to manage their projects. The PMO also provides project managers for different projects, relinquishing responsibility for the result of the project to the PMO.

━━━━━━━━━━━━━━━━━━━━━━━━━━━━━━━━━━ ■ ■ ■

The related steps may include the actual selection of the project or phase. This is when the output of the project goals and objectives, the high-level deliverables, and time and cost estimates to accomplish the business need are determined. The initiating phase also includes creating a narrative description of the project and the responsibilities of the project manager and high-level resources.

During the planning process the scope statement and scope management plan are created. They determine the project team and develop the work breakdown structure (WBS). The WBS is a very important part of any project, but even more so for DVS. There are many intricacies with DVS design and implementation that aren't typical for most other electrical, mechanical, or even network implementations, so the more detailed the task list in the WBS, the better.

The WBS is used to create the project plan. The Project Plan is a comprehensive definition of the project detailed through tasks, resources, and time. A Microsoft Gantt Project chart isn't a project plan; it's just a Gantt chart and only part of the project plan. The project plan includes, but isn't limited to, the following information:

- Project charter
- Major milestones
- Scope statement
- Resource management plan
- WBS
- Change control plan/system
- Flowchart diagram
- Management plan (PM plans)
- Budget and cost management plan
- Procurement plan and procurement management plan
- Schedule and schedule management plan
- Quality management plan
- Risks
- Communications management plan
- Responsibility chart
- Performance measurement baselines

The planning process includes finalization of the team and a resource management plan. Human resources aren't dependable – they might be pulled from one job and placed in another – so it's important to develop a plan to determine how you work with changes to people's schedule or position.

Creating a WBS dictionary includes putting all acronyms, names, and terminology in one place to solidify communications. This is important because in wireless networking a radio can also be a node, station, portal, AP, mesh node, mesh bridge, wireless bridge, or router, depending on where and what equipment was historically used by the resources and suppliers. A wireless portal can also be an entry point, backhaul, or master depending on the manufacturer. They all mean the same thing individually, but may have different meanings to each other.

Planning Process

In my experience with hundreds of projects, there's a simplified methodology that only includes a discover, design, and deploy phase. This only works with smaller projects and clients that don't require documentation and detailed deliverables with the implementation. The larger projects follow a more advanced methodology, including discover, design, document, deploy, and deliverables.

Discover
The discovery step involves gathering the requirements that together detail the ultimate goal the customer would like to achieve and become the foundation for the entire project. There are a few ways to get the requirements to the necessary level of detail to conceptualize an accurate design. This process usually means research, interviews, and

sometimes even digging in the garbage can inside a telco closet that hasn't been dumped in a decade (don't laugh – true story).

A DVS system incorporates a number of elements, all very different from each other, and may include completely different departments and/or personnel. As discussed in Chapter 6, the site survey provides many of the overall details on how the solution may be conceptualized, based on the customer's vision. At times, getting the full view of that vision may feel like pulling teeth, but the more details documented beforehand, the better the chances the project will be successful.

Assumptions

There may be missing details, such as the condition of conduit pathways, power availability, or the condition of existing cameras. These would be listed in a design document as Assumptions. For example, short of rodding and/or blowing the conduit pathways to determine that they're in fact clear and include adequate room for additional power and/or data runs, the usual assumption is that they're clear and available. This is done to save the time and money required for such an undertaking during the discovery stage. If the conduit pathways are blocked or stuffed, then a Change Request is issued to the customer to run new conduit and/or to replace the existing pathways.

Another popular assumption is the availability of uninterrupted power to each specific location where either network equipment or cameras will be installed – power that, in this case, isn't set to a light timer or a photocell and is continuous 24/7 without interruption (different than a UPS backup). Assumptions are always approved by all parties involved and if they prove to be incorrect, a plan of action (including associated cost) needs to be discussed and agreed to beforehand.

The output of this information (discovery phase and related assumptions) is usually a requirements management plan, but I have included it in the design document as part of the why for the conceptual design choices. Because it's all one document, you can reference this section when explaining the various decisions made in the DVS design. For example, during the discovery phase a camera model was determined to be incompatible with the new VMS system and thus needed replacing. The steps taken during the discovery phase (which includes the site surveys) can be referenced to clarify (and remind the client) why this potentially expensive decision was made.

Design

The design phase takes all the accumulated data from various sources (and experience) and develops a DVS solution to achieve the objectives of the project. Design includes making choices on cameras, digital video encoders, network equipment, the VMS software (either existing or new), and archiving and storage hardware. Essentially, once the design phase is complete, a clear bill-of-materials is outlined, and resources, time line, and costs associated with that design are created.

Unless the first stage of the project is an experimental or exploratory pilot phase, it's best to get as much detail as possible to generate an accurate accounting of the implementation costs. This is the point when a high-level WBS is created, with the proposed team members each bringing their expertise and experience to the table.

Figure 9-2 depicts the same data and power flow diagram from Chapter 5 (Figure 5-40). It assumes two different interim distribution facilities (IDF), which can be in the same building, in two different buildings on the same block in a campus environment, or on two different floors. There's also the main distribution facility (MDF), which is considered the main data center, and a new command center. The site survey uncovers the fiber (or wireless) connectivity between the two IDF locations and the MDF; one of the reasons for choosing the IDF locations as well as creating an interim location for terminating the camera is that the MDF is too far for every camera. Either a commitment or an assumption is made that the customer will provide four strains (one spare pair is a good idea) of fiber for the dedicated DVS network between the IDFs and the MDF. Whether these two IDF locations share the same building or the same campus, they're still two completely separate locations that require power, equipment, and resources. The MDF requires a similar installation, but since the VMS

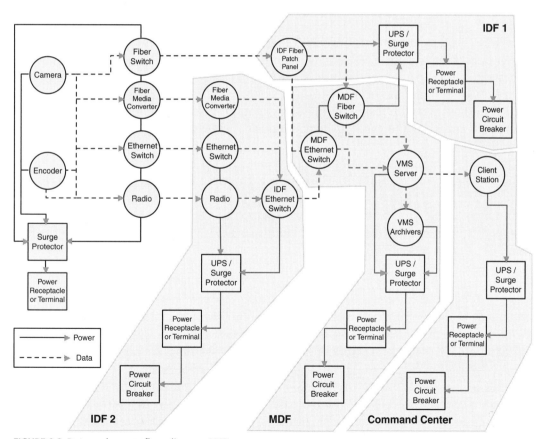

FIGURE 9-2 Data and power flow diagram PMP.

server and network router will most likely be in the MDF, that installation will be more complex. Depending on the schedule, these three locations may require concurrent installations and thus demand three times the resources (or four times the resources if there's a new command center constructed).

Work Breakdown Structure

The WBS dissects all the required work into steps, and those steps into tasks per individual resource, to determine what level of expertise is required, how many resources, what type of equipment is needed, and how many hours it will take to convert the location into a usable DVS IDF.

Table 9-1 is an example of what this part of the overall WBS may look like without noting the added support required to link the DVS network from the IDF to the MDF, and to commission the cameras into the new VMS software in the new command center. This table focuses solely on tasks specific to the IDF, not installation of the new cameras (and it assumes that all equipment was successfully delivered).

Shipments and Delivery

The delivery of expensive and/or sensitive equipment such as video surveillance devices can be a problematic issue, especially if there are UPS backups involved, each of which can weigh about a thousand pounds. It's not an easy task to move these lead acid battery-packed units around without the use of a forklift. This is something that needs to be considered when designing the DVS solution, because it's a problem if the UPS backup is stuck on the loading dock because there's no way of moving it into the building.

It's also important to consider the customer's own shipment security procedures for delivery of the DVS equipment. If working with select corporations, municipals, and government agencies, there may be forms that require completion and delivery before a shipment arrives at the loading dock. Sometimes the company requires details on the shipping company, driver, and truck number before granting access to the loading dock (see Figure 9-3).

The "Eye" in Team

Assigned to the project as early as possible, the project manager is the individual responsible for managing the project and must have the authority and accountability necessary to accomplish the work. A project manager without the authority to fully perform his duties isn't an actual project manager; he's a scapegoat.

A project manager's authority is proactive and includes the power to say "no" when necessary, because he's held accountable for project failure. Project managers, who don't require a technical discipline, also understand professional responsibility; although they're in charge of the project, they may not be in charge of the resources.

Table 9-1 Example WBS for IDF Installation

		Work Breakdown Structure		Project #				
					Sponsor			
Project Manager								
ID	Task	Dependencies (ID#)	Start Date	Estimate to Completion	Resource(s)	Duration	Cost
1	Install new 30 AMP, 120-circuit breaker in each IDF to support new equipment				Electrician		
2	Run conduit from circuit breaker to network rack area				Electrician		
3	Unpack, inspect, and install new UPS Backup and Power Distribution Unit to network equipment rack				Network Administrator		
4	Hardwire power into UPS Backup and Power Distribution Panel				Electrician		
5	Unpack, inspect, and stage networking equipment, and Ethernet and/or fiber-optic patch panels in the intended networking room (IDF)				Network Engineer		
6	Position the networking devices in the equipment racks per the prepared rack elevation design drawings for each IDF				Network Engineer		
7	Connect AC and DC power supplies to optical networking equipment, and check for successful POST routines	1			Network Engineer		
8	Run initial diagnostics to affirm correct operation of networking devices				Network Engineer		
9	Unpack, inspect, and stage digital video encoders in the IDF				Network Engineer		
10	Connect position and configure the two network switches and patch panels in the equipment racks per the design drawings				Network Engineer		
11	Run initial diagnostics to affirm correct operation and document for Final Acceptance Test Plan						
12	Determine single mode fiber pairs in the IDF closet				Telecom Engineer		
13	Connect network switch in IDF to fiber pairs identified and provided by customer				Telecom Engineer		
14	Install SFP connectors in IDF switch				Telecom Engineer		
15	Connect fiber jumper to identified and provided fiber pairs				Telecom Engineer		

Continued

Table 9-1   Example WBS for IDF Installation—Cont'd

Project Manager				Project #			
					Sponsor		
ID	Task	Dependencies (ID#)	Start Date	Estimate to Completion	Resource(s)	Duration	Cost
16	Power level test (only) will be performed on the fiber-optic cabling between the IDF and the MDF and document for Final Acceptance Test Plan				Telecom Engineer		
17	Hardwire the Ethernet cabling from the 50 new IP cameras to the patch panels and use the Ethernet jumpers to connect to the network switches				Network Engineer		
18	Install two 12 input port digital video encoders onto network equipment rack per rack elevation design document				DVS Engineer		
19	Disconnect coaxial cables from existing surveillance cameras that have been deemed operational through site survey testing and connect and/or add coaxial extension to connect coaxial cable from exiting surveillance cameras to digital video encoders				DVS Engineer		
20	Connect PTZ control wiring to the related 12 serial ports in digital video encoders				DVS Engineer		
21	Using the Web interface, configure digital video encoders with assigned IP addresses, subnet mask, and gateway interface and configure PTZ control of serial ports and test video in live viewer; document for Final Acceptance Test Plan				DVS Engineer		
22	Update as-built documents with any deviations from the rack elevation design document and document for Final Acceptance Test Plan				DVS Engineer		
23	Test digital video encoder connectivity and document for Final Acceptance Test Plan				DVS Engineer		
24	Configure IP routing devices with IP addressing scheme to conform to the design documents and document for Final Acceptance Test Plan				Network Engineer		
25	Perform network acceptance testing to verify the network functions as designed in the design documents and document for Final Acceptance Test Plan				Network Engineer/DVS Engineer		
26	Complete post-installation quality audit and document for Final Acceptance Test Plan				Network Engineer/DVS Engineer		

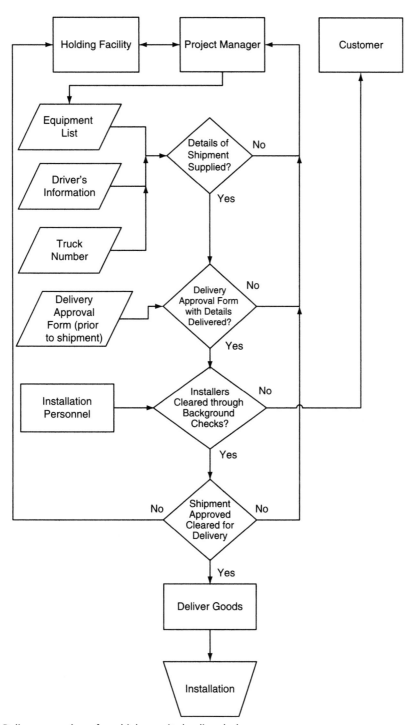

FIGURE 9-3 Delivery procedures for a high-security loading dock.

The Triple Constraint

It's very important to understand that a project manager is responsible for the overall efficiency, schedule, and execution of the project but not necessarily the individual tasks. Whether projects are small or enterprise in size, there's one constant in all of them – continuous juggling of the "triple constraint." This is the balance of the project's triangle of scope, schedule (time), and cost (Figure 9-4). It measures the successful delivery of the project's objectives. As part of the development of the project plan, the project manager and/or project management team defines the project scope, time, cost, and quality of a project. As the process continues, and some changes or adjustments are discovered against the project's scope, time, or cost, the triple constraint is usually affected. For example, if the schedule is expedited, more resources are required (cost) and the scope will change, and if the cost decreases (funds are cut), then the scope and time will also be reduced to compensate for the cost reduction.

FIGURE 9-4 Triple constraint – the continuous balancing of scope, time (schedule), and cost.

The project manager is there to juggle the ever-competing demands for scope, time, cost, and quality, and communicate that balancing act to the project stakeholders with the changing needs and expectations. Control of the project and its assets helps to prevent scope creep, which is the gradual increase of project costs, time, and scope, and usually occurs when the customer asks for additional work outside the scope of the statement of work.

A project manager is much more valuable monitoring the overall components than actually executing them (although he may be responsible for the actual project plan). The project manager manages. However, there may be exceptions based on the size of the project and/or organization.

In my youth, as an entrepreneur, I experienced a time when my business outgrew me physically, mentally, financially, and emotionally. There was no more blood and sweat to give and time became my most valuable commodity. There wasn't enough time in the day to handle everything that required my attention. I believe there's a line that every entrepreneurial company crosses during its growth and three other entrepreneurial adventures in the 1990s validated my theory. This line symbolizes the moment the entrepreneurial entity struggles to become a corporation; when the company, as the overall project, becomes too cumbersome to manage without more structure and to create more structure all hats but one need be present. In an entrepreneurial environment shortcuts are taken by everyone involved and everyone wears many hats. In a small company, it may work well for everyone to wear many hats, thriving on teamwork until the wee hours of the night. However, as a company grows, a handful of people blossoms into several dozen people, each with a unique task that adds to the complexity of the project (and/or projects) and its goals. The entrepreneur, who's typically the glue that holds the team together, finds that with a larger circle of people and an exponential growth in tasks, schedules, milestones, and deadlines, there's no time to assist in the detailed execution because managing the project and/or subprojects now takes precedence. The project manager is like the entrepreneur who becomes the glue that holds the pieces together.

Shortly after I wrote my first business plan and acquired a couple of investors, one of them entered my office and asked me, "What exactly do you do around here?" I laughed at this question, in part because I thought it an odd question coming from someone who was four times my senior, so I told him, "I'm the glue." It was shortly before this incident that I had come to understand that when managing a large number of people it's important to avoid becoming the answer man. Don't make it easy for associates (and/or employees) to just come up to you and ask a question, relying on you to have an instant answer. I found myself just sitting in my office, avoiding beginning anything because I knew that within 3 to 5 minutes (yes, I actually timed it once) I would be interrupted by someone with an urgent question. This may be unavoidable at the start of a project, especially when training new recruits, but it's best to promote thinking, not just doing, with the question "What do you think you should do?" This is especially true for delivery project managers – those who have been assigned a subproject due to a unique skill set, time line, location, or milestone.

I shared this story with my students when I was a PMI certification instructor. PMI believes in a strict project management methodology, which is a challenge to follow in the real world. Many of my students had been project managers for decades and the PMP certification was the next step in their career path. I taught them project management, as defined by PMI, for the benefit of passing the PMP exam. Once my students understood that, there was much less debate on the subject.

In the real world a project may run out of funds early, thus chopping the closing procedures and leaving the result completely devoid of documentation on lessons learned. Also, in the real world there's less time for detailed deliverables including documentation, as-builts, and even training. Unavoidable disasters happen, assumptions are underestimated, resources come and go, and the list goes on. However, the PMI methodology is superior to anything else out there and especially with the multilayered complexity of a DVS project, detailed structure creates a stronger foundation upon which to build. The key is foreseeing how the project may be impacted by these potential problems and planning accordingly.

The planning process is also the time to gather the information required to create the network diagrams. In project management, the network diagram is a high-level flow chart of the processes of the DVS system implementation. Microsoft Project provides a network diagram view of the project plan.

Critical Path Method

The planning process also includes the financial and scheduling evaluation of the WBS as well as estimating the time and cost of the project, using the Critical Path Method (CPM) for scheduling and project planning. This method is the absolute scheduled beginning and end of a project. If any activity within this path is delayed, it delays the entire project. The project manager monitors the overall project because, as the Gantt chart illustrates, there are layers upon layers of tasks typically in place and some can be shuffled earlier or later, or groups or phases moved around in the schedule, while others require preceding milestones be met before execution. Somewhere in the project plan there's an absolute path from beginning to end. If one task or phase includes a requirement that the preceding milestone be met in a timely matter, delay of that milestone means the entire project will be delayed. For example, installing a camera in a location still waiting for power may delay the entire project if there's a problem with delivering power to that location or it becomes even more costly than originally estimated. Wouldn't it be worth waiting to determine if that's the best location for the camera or if 25 feet to the east would be better? Figure 9-5 depicts the critical path of a high-level DVS implementation. This critical path was determined using the example in Table 9-2.

In addition to the duration of the tasks changing the schedule, there's also the possibility of an early or late start and finish. It's important to evaluate any flexibility in the task's schedule related to the other tasks before, concurrently, or after.

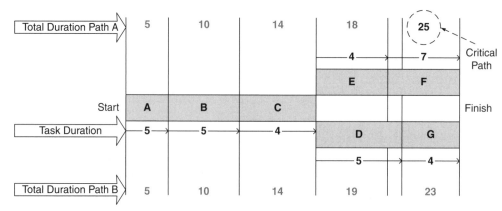

FIGURE 9-5 Critical path.

Table 9-2   CPM Preceding Task Relations

Task	Preceding Task	Duration
A	[Start]	5
B	A	5
C	A	4
D	B, C	5
E	D	4
F	E	7
G	D	4

Quality Management

Quality is defined as the degree to which the project fulfills requirements and gains the customer's satisfaction. Quality management is a management function that determines quality policy, objectives, and responsibilities. This is accomplished through quality planning, quality assurance (QA), quality control (QC), and quality improvement within the system (see Table 9-3).

A QC plan is a documented, comprehensive set of procedures and activities aimed at delivering contractual requirements to meet or exceed expectations. The QC plan identifies those responsible for QC and the related procedures. QC consists of specific operational activities put in place to manage the quality of the incremental tasks and outputs of the project. These include providing clear decisions and directions, supervision and mentoring by experienced individuals, and review of completed tasks for accuracy with accurate documentation. QC procedures raise the level of quality services and products provided without micromanaging each task and resource.

QA refers to the certainty that the incremental tasks and outputs meet the requirements and to the continual improvement of the total delivery process to enhance quality, productivity, and customer satisfaction. Essentially, QA describes the process of enforcing QC standards. When QA is well implemented, errors and omissions are reduced and increased usability and performance are achieved. QA is the voice of the customer.

Table 9-3   Quality Management Planning, Assurance, and Control

Quality Planning	Quality Assurance	Quality Control
Implement existing quality standards for product and project management	Perform continuous improvement	Measure specific project results against quality standards
Include additional project-specific standards	Determine project activity compliance with organization and project policies, processes, and procedures – quality audit	Implement approved changes to quality baseline
Establish daily internal status meetings and weekly stakeholder meetings	Identify and correct deficiencies	Identify quality improvements
		Repair defects
Balance quality with the other triple constraints	Identify improvements	Recommend changes, corrective and preventive actions, and defect repair to integrated change control
Create Quality Management Plan and add it to the project management plan	Recommend changes and corrective actions to integrated change control	
	Hold daily internal status meetings and weekly stakeholder meetings	

Process Improvement

An important aspect of project management is identifying the need for improvement in the midst of chaos. Process improvement (see Figure 9-6) is successful when an effective process emerges or evolves that can be characterized as practiced, documented, maintained, trained, measured, and improvable.

A general process improvement program includes the detection of variances in procedures along with the identification of responsibility. An evaluation should identify the importance of

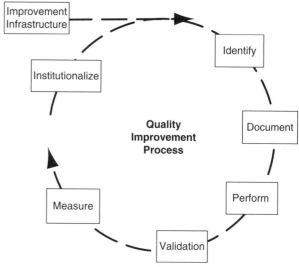

FIGURE 9-6 Process improvement.

the problem and an investigation of possible causes determines the severity of the problem as well as any preventative action required to continue QC.

Risk Management

The definition of risk as it relates to project management is a potential event or future situation that may adversely or positively affect the project. The characteristic uncertainty of risk distinguishes it from an issue, because risk addresses a potential event while an issue is a known concern. The risk management plan identifies potential problems that may arise in the course of the project execution and how those problems will be addressed.

An analysis of risk becomes clear when researching, analyzing, and developing the project requirements. Project risks are risks to the business objectives/critical success factors that in the project manager's judgment are sufficiently serious to require attention. Risk management is the continuous identification, reporting, analysis, prioritization, monitoring, and control of internal and external activities, conditions, and/or events that can cause an undesirable project impact.

The risk management process has three major steps:

1. Risk identification
2. Risk quantification
3. Risk response and control

It's important to note that the project manager facilitates the risk management process and may invite other stakeholders to participate, because some risks are unique to a customer's business and may not be identifiable or addressable by the project manager or the PMO (Risk assessment process flow in Figure 9-7).

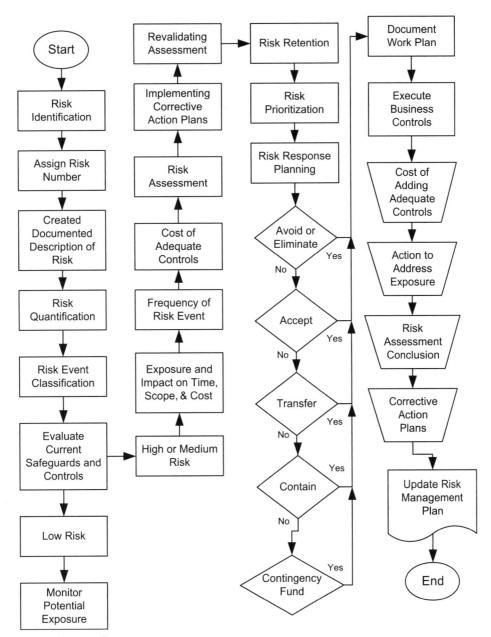

FIGURE 9-7 Risk process flow.

Risk identification begins at contract negotiations and continues throughout the life cycle of the project. Typical triggers include the start of the project itself, key events or milestones, and any changes in the environment or the business objectives. Other triggers might be key technical changes such as new technology or unavailability of a specific technology and changes in the statement of work. Once a set of risk events has been identified, the project management team does some preliminary prioritization. They may need to collect additional information to facilitate the next step of evaluation.

In any complex implementation, there are a large number of risk factors. A clear evaluation process involves accumulating critical information to judge the severity of the risk. As part of the evaluation, each risk should be analyzed both on the probability of the event occurring and the impact to the overall project, if or when the event does occur. Project management should set the criteria for each contract. Once the management team agrees to the criteria, the risk events can be analyzed and the amount of contract resources to be focused on an individual event decided.

If a risk is rated as an acceptable response (the stakeholders can live with the results of this risk), the risk is simply identified and acknowledged with no further action. Any other risk should be identified and monitored with the execution of a response as part of project management. Risks should be proactively managed as part of the daily activities based on the information collected.

Risks can also be positive, such as a new business development arising from potential changes in the business goals of the customer, changes in the customer management, or the introduction of new technology.

Risk responses may include avoidance or complete elimination of the risk, such as choosing an unproven camera technology. Perhaps the potential is enormous and the technology way ahead of its time. But do you want your project to be the beta site? Once risks are identified, there may be acceptable risks that only require documentation of the consequences, should the risk occur. For example, if it rains or snows, the rooftop installation will be delayed due to union rules. Transferring all or part of the risk may include delegation of a task or subproject to a subcontractor, who then assumes all of the risk for that portion of the project. Containing the risk through specific actions to lower the probability and/or impact includes possibly isolating that portion of the project and breaking it down into a smaller piece as a pilot program. This would provide accurate data on the outcome and impact of the risk, if it occurs.

One of the best courses of action for risk management is setting up a contingency fund with the sole purpose of providing support for any unforeseen events that may occur during the course of the project.

Risk tracking and control always depend on the availability of accurate, relevant, and timely data. Data related to individual risks need to be collected, compiled, and analyzed to determine what actions should be taken (and taking no action is considered an action). Risk tracking and control include:

- Collecting tracking data related to risks
- Analyzing data related to individual risks
- Deciding risk response actions based on risk data
- Closing the risk when appropriate

Initially, risk status should only be communicated internally, on a regular basis, via a status reporting method and meetings before presenting it to the customer.

■ ■ ■ ──

Contract Deliverables Requirement List

A Contract Deliverables Requirement List (CDRL, pronounced "SEE-DRILL") is a tool that provides a method of managing the specific requirements in the contract and/or comments to a document deliverable. The CDRL (see Table 9-4) is a spreadsheet and contains each point addressed by the customer and/or PMO, where it may reside inside the contract and statement of work paragraph reference, and whether that task was assigned and accepted.

The CDRL is in accordance with the following column headings:

- CDRL#: Item Number
- Section Number: Reference Section Number
- Task Description: Deliverable Description
- Due Date: Scheduled Delivery Date(s)
- Assigned To: Who Owns the Action?
- Acceptance: Acceptance Status (i.e., pending, approved, conditionally approved, disapproved)
- Notes: Comments

Table 9-4   CDRL Example

CDRL#	Section Number	Task Description	Due Date	Action Status	Assigned To	Acceptance	Notes

── ■ ■ ■

The Project Plan

The project plan is a presentation of the time (schedule), scope (tasks), and costs (resources) of the implementation. Many may feel it necessary to purchase project planning software to help with the creation of the project plan, but this software doesn't create a project plan (not even with Microsoft Project). These are tools used to accumulate, organize, and eventually present the project plan.

The project plan is different for every project because no two projects are exactly the same. As much as we would all like to have a simple template to plug in to each project as it arises, such a wish is unrealistic. The project plan must be unique to each project, driven by the requirements and design. It may be more than one document. The project plan for large implementations may require separate communication, risk, quality, and resource management plans. The actual project plan may have an executive summary of each management plan with the full document added as an Appendix.

The major component of the project plan is the integration of all the various elements into a project schedule with resources and milestones. Milestones are an important part of any project. One reason is the forward momentum achieved by carefully managing the preceding and subsequent milestones laid out along the critical path. These milestones are crucial to the project schedule and can be postponed by an unknown impediment in a preceding milestone, which at first glance may not seem vital. The second reason is payment. Many projects have payments linked to meeting specific milestones, so choose an accurate assessment of its completion date by considering historical information from any resources assigned to assume responsibility for the task, including any outside project stakeholders (such as subcontractors).

Project Stakeholders

A project stakeholder is anyone whose interests are positively or negatively impacted by the project and who may exert influence over it. Identify their needs, expectations, and objectives and determine the roles, knowledge, and skill set of each stakeholder on the project. It's also vital to include them in the planning process, taking in their historical knowledge from previous projects.

Sponsor

A sponsor is the person who provides the financial resources for the project, which in most instances is also the customer, but can be a project management representing an organization or an individual whose role as the sponsor is to formally accept the product of the project during scope verification and administrative closure. Along with the customer and stakeholders, the sponsor may provide key events, milestones, and deliverable due dates along with the threshold for risks. The sponsor isn't the individual who signs a project charter or Statement of Work (SOW); that's the role of senior management.

Project Life Cycle

A life cycle is a progression through a series of different stages of project development.

A project life cycle describes how the work involved in the project is completed and is specific to the technical disciplines, what technical work should be done in each phase, and the project deliverables for each phase (Figure 9-8).

The life cycle includes the project resources and in which phase those resources would be involved as well as how to control and approve each phase. Cost and staffing are low at the start but higher toward the end as milestones are reached and more resources begin new tasks. The probability of successfully completing the project is lowest at the start of the project, whereas how the stakeholders influence the final characteristics of the project is highest at the start of the project.

The project management processes represent the whole life of a project except when a project is broken down into multiple projects (sometimes called a "program"), which the process groups may repeat throughout the project life per each stage or phase.

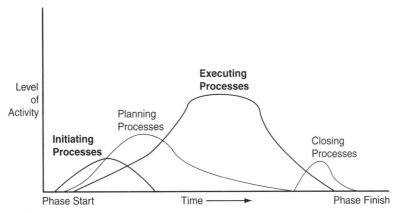

FIGURE 9-8 Process flow.

Execution Process

Once the project moves into the execution process (phase), usually after a project launch or kick-off meeting, the control and monitoring kick into high gear. The project requirements should be detailed enough to track progress for the status reports. Project tracking can be done at the task level, which is the least detailed level, because the task is the only work updated and reported. The report includes the total labor required to complete a task and the amount of work a specific resource is assigned. Work isn't the same as task duration, which is a projected time for the task to be completed, not the actual time. An assignment level provides more tracking details by monitoring tasks that have one or more resources assigned to them, each associated with work and cost values. Progress tracking can also be done as totals, which is aggregate cost or work of a task or assignment up to the current date or cost. For example, what's reported is that a resource has worked 25 hours on an assignment, which can also contribute to tracking progress by time period to focus on staying on schedule or on budget.

Project Tracking
During the execution phase the project manager begins to track the progress of the implementation. He establishes an escalation procedure (see Figure 9-9) for emergencies – a change control process in the event that something goes awry from the project plan and system design – and coordinates subcontractors. To accomplish this, the project manager must be in constant communication with all project personnel and/or their supervisors. The information gathered is critical to evaluate progress against the baseline of the plan and to prepare for status meetings and reports. It's also usually the project manager's responsibility to communicate with billing on the accounts receivables and payables unique to the project.

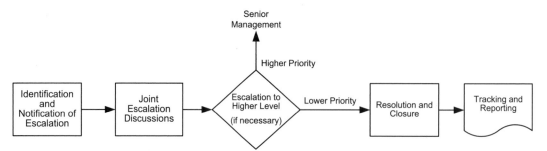

FIGURE 9-9 Knowing the chain of command can expedite emergency decisions.

TRACKING USING MICROSOFT PROJECT

Depending on the details you decide to track and the size of the project, Microsoft Project provides various methods of tracking the progress of a project. The task-total method tracks the total task durations, work, and/or costs to the current date, milestone, or status date. In Microsoft Project, use the Task Sheet view with the Tracking table displayed by pointing to Table on the View menu and then clicking Tracking.

The task-time-phased method tracks task work and/or costs per time period by using the Task Usage view with the Actual Work details by right-clicking the time-phased portion of the Task Usage view and then clicking Actual Work.

The assignment-total method tracks the total work and/or costs per resource assignment up to the selected date. In Microsoft Project, use the Task Usage view with the Tracking table displayed.

The assignment-time-phased method tracks each resource assignment's work or costs per time period by using the Task Usage view with the Actual Work details attached. The amount of detail presented in a status report (either in written or verbal form) depends on the comfort level of the customer. Typically, the larger the project, the more detail the customer requires; more detail is also required for control and monitoring. Without a high level of detail, it's difficult to keep a handle on a project as complex as a DVS implementation.

Status Meetings and Status Reports

Project status meetings and reporting are visible activities that prove the progress of the implementation. Depending on the size of the DVS implementation, the customer typically likes to know how the project is progressing, and not just to gain proof of how the money is being spent.

Part of the status reporting includes receipt of equipment deliveries, schedules, and if any of the equipment is back-ordered. The project manager needs to estimate when all the equipment will be delivered so he can hire the resources to install that equipment. Equipment delays can push the schedule back and it's probably a good idea to let the customer know that it's not due to anything within your control. In the meantime, the project manager must work magic to shuffle the schedule around to keep the project moving forward.

The beginning of the execution process is usually where the most status meetings and reporting occur, in part due to the equipment deliveries and the confirmation that the implementation is moving forward without many obstacles, and of those obstacles that do arise, which require the customer's influence.

Change Control Process

The change control process provides the customer with the ability to make changes in the scope of work. It's vital to any project to have a means of controlling any changes (known as scope creep). As a project manager, and through the control and monitoring process, you must recognize that any work beyond what was originally designed devours time and resources from the actual work scheduled to be done and/or the goods. This creates a deficit in the cost and schedule. Changes to the scope include:

1. Material change to the services
2. Increase or decrease in the amount or number of the goods
3. The type of DVS software
4. A change required in the services, goods, system software, or the system itself caused by material unforeseen conditions that require a change in the terms and conditions of the agreement
5. A change in the system, goods, or services that's necessitated by a change in applicable laws or city policies
6. A change to the services, goods, or the system deemed to be potentially beneficial to the customer
7. Deviation of an assumption, which has a direct impact on the services or as otherwise expressly provided in the statement of work

A change request by the project manager should always be in writing, either as a letter, as a written proposal, and/or a change request form. The stakeholders involved and directly linked to the change request will either reject or accept the change (in writing) within a designated amount of time. A deadline directly connected to the project schedule is important to avoid excessive delays.

In the event the change request is rejected, the written response would include the reason for rejection (for future reference) and if applicable, a viable, good faith alternative to the requested change. When the change request is accepted, all stakeholders involved mutually agree to the changes in writing as an official change order.

Except for emergency changes, which are changes that require immediate attention due to project scheduling or procurement, all changes or additional terms and conditions should be made only through a written change order signed by the authorized stakeholders. If there are additional costs associated with the change (or reduced costs), the terms and conditions should either follow the original agreements or be negotiated in good faith to equitably adjust the cost based on the existing project agreement.

Project Documentation Deliverables

Status reports are part of the project's document deliverables. Documentation is an essential part of any project and becomes even more important the more complex the project. Unfortunately, I've been thrown into integration situations where there were no documents, diagrams, as-builts, or even manuals for the existing equipment. I had to assume this was a scheduling (budgetary) issue and not a purposeful attempt to conceal information. At that point, reverse engineering was the only means of understanding how the proprietary system worked and if there was any integration potential.

I believe (and it's not because I'm an author) that document deliverables are important to all projects because they show the customer all the work that was accomplished for their money. Rarely does a customer representative watch each stage or task of the project. Document deliverables record many of those monumental obstacles that were cleared and the numerous decisions, workarounds, and typical issues that are part of any system implementation. There isn't that much time available for someone who already has his day full of regular duties and responsibilities to communicate every little decision. The document deliverables show the customers that they're getting their money's worth.

The types of document deliverables vary dramatically and are solely based on the specific requirements of the implementation, but it's best practice to leave (at the very least) a cheat sheet of how to use the VMS software. This way, anyone, new or old, can learn how to use the very basic features of the DVS system (if authorized).

Table 9-5 provides an overview of the types of documents that may be required as deliverables, depending on the organization and size of the project.

Table 9-6 shows several actual documents produced for a number of DVS projects, their descriptions, inputs, and the tools used to create the documents. The use of these documents as part of a DVS implementation is entirely based on the requirements and size of the installation. The larger the size of the install, the more project management is required to control and monitor the progress.

As-Built Documents

The as-builts documents include the actual recorded implementation of the cameras and network equipment. These may look similar to the site survey forms, but they include the actual location of the camera, not its projected location. During the site survey (see Chapter 6), there may have been a primary and secondary location chosen per camera, based on the availability of power and data, height, and/or the position to the required area of coverage. The DVS design document presents the overall solution and equipment required, but like any system integration and implementation, it's part visionary, part fact, part fortune-telling, and part science. However, the design document is very detailed, but it remains conceptual and thus the final as-builts provide the customer with the real world installation data.

Table 9-5   Project Documentation and Descriptions

Type	Description
Administrative Documentation	Documents pertaining to the administrative operations of the project, including documents for funding, personnel, staffing, equipment licenses, warranties, etc.
Analyses and Recommendations	Documents describing a specific problem or scenario and the anticipated impact and/or recommended course(s) of action.
Contract Management Documentation	Documents associated with the solicitation, administration, and management of the contractors supporting the project.
Correspondence and Communications	Documents sent to or received from any organization external to the project, including the sponsor, control agencies, federal stakeholders, counties, advocates, and the public.
E-mail	Critical e-mail is retained such as important information received from contractors or other outside sources related to the project. Project staff shouldn't use e-mail for formal communication or decision making on the project (as e-mail is a binding contract in a court of law). Any critical e-mail should be archived and non-critical e-mail purged at the user's discretion.
Plans and Processes	Documents describing the purpose and approach to the project, including the plans and processes that describe how the project will be executed and managed.
Presentations	Documents used in training or briefing project staff, county staff, or stakeholders.
Reference Materials	Documents generated by an external organization that provide insight, guidance, or examples of pertinent information such as legislation, policy, regulations, handbooks, standards, etc.
Status Documentation	Documents describing the current status of planned and actual activities for the project including funding, contract, schedule, issue and risk status, and meeting minutes describing decisions, action items, and concerns.
Working Papers	Early drafts, notes, or reference materials used to create another document. Working papers may or may not be retained at the author's discretion.

Table 9-6   Project Document Deliverables

Deliverable	Description	Inputs	Tools
Design and Configuration Plan	A document that includes the overall design concepts and the individual configurations of the proposed equipment	DVS Requirements Document, statement of work from all stakeholders, all site survey forms and documentation, and conceptual design documentation	Microsoft Word, Visio Professional
Implementation Plan	A documented approach to the delivery of the DVS system	DVS Requirements Document, statement of work from all stakeholders, design documentation	Microsoft Word, Visio Professional
Detailed Project Plan	A documented task list, calendar timetable (with associated costs for internal use) for completion (produced in Microsoft Project)	DVS Requirements Document, statement of work from all stakeholders, WBS, all site survey forms and documentation, and design documentation	Microsoft Project Professional and Word

Continued

Table 9-6 Project Document Deliverables—Cont'd

Deliverable	Description	Inputs	Tools
Communication Management Plan	A documented approach to networking maintainability	DVS Requirements Document, statement of work from all stakeholders, WBS, all site survey forms and documentation, and design documentation	Microsoft Project, Visio Professional, and Word
Quality Assurance Plan	A plan outlining the quality philosophy of the project, its subcontractors, and a detailed approach to insuring contractual requirements are met	DVS Requirements Document, statement of work from all stakeholders, WBS, all site survey forms and documentation, and design documentation	Microsoft Project, Visio Professional, and Word
Risk Management Plan	An account of how risks will be recognized, confirmed, documented, and handled	Detailed Project Plan, DVS Requirements Document, statement of work from all stakeholders	Microsoft Project, Visio Professional, and Word
Project Team	An organizational chart of all stakeholders, their positions, duties, and responsibilities	Project Plan, DVS Requirements Document, Quality Assurance Plan, Risk Management Plan	Microsoft Project, Visio Professional, and Word
"As Built" System Documentation	Detailed documented evidence of the actual installation/integration	DVS Requirements Document, statement of work from all stakeholders, all site survey forms and documentation, and conceptual design documentation	Microsoft Office Suite
Final Acceptance Test Plan	First Article Testing procedures confirming system is in operational mode at system launch	DVS Requirements Document, Quality Assurance Plan, statement of work from all stakeholders, all site survey forms and documentation, Risk Management Plan, Project Plan, conceptual design documentation, as-builts	Microsoft Word, Excel and PowerPoint (if a separate presentation)

For camera locations, the as-builts include the location of the IDF (where the camera is plugged into the network) and whether it's an IP camera plugged into a network switch in the IDF, if it uses a fiber transceiver that converts Ethernet-to-fiber for those runs longer than 300 feet (100 m), or if the camera is an old world analog camera hooked up to the digital video encoder in that IDF. The as-builts would also include actual port numbers on the switch and/or digital video encoder, a mapped location (if it's a large implementation), housing and mount type, and a photo of the actual camera installed and screen captures (usually north, south, east, and west if it's a PTZ camera) of the camera's area of coverage.

For the network locations, the as-builts for the IDF include the actual rack elevation diagram (use Microsoft Visio to create these), power specifications, and a list of cameras being serviced. The MDF also requires the VMS server and the storage area network RAID array configuration.

The command center requires a system installation diagram, depending on the amount of equipment set up for the DVS system. A single workstation, with at least dual video graphics cards, requires a system diagram to allow the administrator to trouble-shoot the system in the event of failure.

I've found that sometimes a good diagram is acceptable as an as-built, depending on the detail. This saves time writing a documented account of the installation with a nar-rative description that no one reads. It's easier to take a look at an exceptional diagram, flowchart, or graph to absorb the overall configuration than to find the time to sit down and read a narrative about the system. However, the choice to substitute a diagram really depends on the customer requirements, but this could be presented as an optional solution.

Drawings, Schematics, Charts, and Graphs

There are two standard types of image files. I'm not talking about the image format, but its core substance. There are rasterized images, which are basically images that were scanned or created using pixels per inch. The Internet is full of rasterized images in JPEG, GIF, and PNG formats. For higher quality print TIFF and/or EPS formats are used. The difference between the two is the compression. JPEG is a lossy compression – it dumps data to make the file smaller for viewing on the Internet – but the TIFF format keeps all the data. Obviously, the TIFF file, even at the same resolution, would be considerably larger and thus is used for printing.

As explained in Chapter 2, digital imagery on the computer screen (including the Internet, software, DVDs and DVS) is at either 72 or 96 dpi (dots per inch). Print, on the other hand, requires 300 dpi for the best quality. This can make document deliverables too large to manage (and Microsoft Word can only handle so many). Vector Graphics, which are limited to diagrams and flat or line art only, aren't rasterized images. They're made up of pixels, but are built differently. This is why they're considerably smaller in size and better quality.

Figure 9-10 depicts two images at two different sizes: standard size and the same image enlarged 500%. The top images are photographic, and the bottom are vector graphic images. The differences are quite dramatic. Software applications such as Adobe

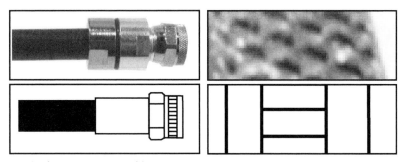

FIGURE 9-10 Rasterized versus vector graphics.

Illustrator, Computer-Aided Design (CAD), and Microsoft Visio all create imagery in vector graphic form (similar to the center images). They can also be saved as vector graphics (typically as EPS or SVG [Illustrator], VSD or SVG [Visio], or DWG or DXF [AutoCAD]) and many can still be inputted into a document. Vector graphics have another advantage. If the drawings are incorporated into documents, the original lines and letters can be emboldened to retain their legibility after reduction. All reproductions should be of high quality, with no blotched or poorly readable areas. One poor-quality image can reflect on the entire document.

All document deliverables need to follow common categorization, symbols, and designations that are approved by all associated bodies, usually in writing, unless otherwise noted inside the document's definition section.

Version Status and Numbering

It's important, especially with multiple contributors, that all required documentation include document data specifics and are numbered to reflect current changes to the document.

The document information should include the version number, document name, and file name (if different). Many organizations use a date in front of the file name to manage versioning (which works until two people make changes to the document on the same day). When a major revision changes the document significantly, the version should move to the next numeric identification (e.g., 1.0-2.0). If there are insignificant changes, such as expanded or edited sections, corrected typos, and new charts and graphs added to existing chapters or sections, then changing the number of the second octet in the official version, such as 1.1 or 2.1, indicates a revision.

It's recommended that documents requiring an iterative approach should remain as a draft until the select phase of the documented project closes and the document is considered complete.

Example Tables for Document Deliverables

Below are examples of tables that can be used to track document deliverables.

Version number	2.0
Document name	Document and Procedural Standards v2.0
File name	10102010_DocumentProceduralStandards_v2.doc
Repository/Access	

Name	Function	Approval Date
Approvers		
Customer Approvers		

Revision STATUS

Version	Date	Changes
1.0	XX/XX/XXX	Preliminary draft
2.0	XX/XX/XXX	Description of revision

Prepared By: [Author's Name]   _____   _____
                                           Signature   Date

Reviewed By: [Project Manager]   _____   _____
                                           Signature   Date

A revision status block should be provided for all submitted documents including maps, diagrams, floor plans, and conduit pathway plans. The revision block should identify the revision number/letter, date of revision, a description of the change, the creator of the document, and the person adding the revision change.

Closing Process

Final Acceptance Test Plan

The Final Acceptance Test Plan validates and documents that the as-built system is performing to design specifications and has successfully fulfilled requirements. The very nature of the DVS implementation requires both modular and system-wide testing models. This minimizes time and expenses associated with testing the system, and shows that during implementation, testing of the modules was accomplished to commission the cameras into the new VMS system.

This process ties into the tail end of the commissioning process, which is the last stage of the installation of a camera and all its required support components. These support components include fiber infrastructure, network equipment and camera equipment, digital video encoders, servers, and storage and management equipment. The Final Acceptance Test Plan assumes that all of the support components have been installed and tested, and the documentation is available to review such as the network topology diagram, as-built diagrams, drawings and photos, and the functional requirements.

MODULAR TESTING

Modular testing consists of operating the components independent of the system as a whole. The components of any DVS system include cameras, digital video encoders, radios, network switches, and power supplies.

CAMERA TESTING The camera test and configuration should be performed prior to the installation of the camera, especially if that camera is going to be installed 25 feet high and out of reach without a bucket truck. This involves configuring and installing the camera into an outdoor housing (if applicable), adding the PTZ control printed circuit board into the head of the camera or housing (if not part of the housing), and then wiring and testing for power and communications.

These are pass or fail tests and documentation is unnecessary (did I say that?). If the camera doesn't pass the test it should be sent back to the manufacturer for a replacement. If the camera passes the initial tests, then it's added into the DVS network where it will be tested once again as part of the system as a whole.

The camera tests should include the following functional tests:

1. Unpack and visually inspect the camera for any damage during shipment.
2. Inspect compatibility with installation components and parts.
3. On a static-free powered workstation, connect the camera to the temporary power source and the PTZ printed circuit board (if applicable) to confirm initial operation and power-up test cycle, verifying normal startup operation in accordance with the product manual.
4. If an analog PTZ camera, ensure that all dip switches are set to the right configuration for the digital video encoder.
5. Far Vision Test (PTZ): Focus the camera on an object at a distance of 300 feet with full zoom.
6. Near Vision Test (PTZ): Focus on an object at a 30 foot distance from the camera lens.

Many cameras run a power-on self-test (POST) boot cycle. This is done with IP cameras, which typically use some version of Linux as an embedded operating system, so each camera can validate itself by running onboard self-diagnostics and provide either an onscreen message or a log file.

DIGITAL VIDEO ENCODER TESTING These are also pass or fail tests. Remember, this isn't about troubleshooting. The Final Acceptance Test Plan is the process of gaining acceptance of the system by the customer. All the equipment should be new and under manufacturer's warranty. If the digital video encoder doesn't pass the test, it's returned for a replacement unit.

The digital video encoder tests include the following functional tests:

1. Unpack and visually inspect the digital video encoder for any damage during shipment.
2. Inspect compatibility with installation components and parts (make sure the right power adapter is available).
3. On a static-free powered workstation, connect the digital video encoder to the temporary power source to confirm initial operation and POST cycle.
4. Connect an analog camera to the digital video encoder and configure to access the video to confirm proper operations.
5. If applicable, connect the PTZ serial controls to confirm the PTZ works with the component camera.

NETWORK SWITCH TESTING Again, testing new equipment right out of the box is either pass or fail. If the network switch doesn't pass the simple test below, it should be replaced.

1. Unpack and visually inspect network switch for any damage during shipment.
2. Inspect compatibility with installation components and parts (is the power cord included?).
3. On a static-free powered workstation, connect the network switch to the temporary power source to confirm initial operation and POST cycle.
4. Plug a Cat5 Ethernet cable into each port (while connected to either a laptop or another powered switch) to confirm port operations through the illuminated LED on the face of the switch.

Number four is a good way to test existing network switches. If an existing switch is used, make sure that all ports are operational and then test for connectivity. If even a single port is dead, best practice is to replace the entire switch (as a change order).

WIRELESS RADIO TESTING If the wireless mesh radio doesn't pass the simple tests below, it's defective and should be returned for a replacement unit.

1. Unpack and visually inspect the wireless mesh radio for any damage during shipment.
2. Inspect compatibility with installation components and parts (is the power adapter included?).
3. On a static-free powered workstation, connect at least two wireless mesh radios to the temporary power source to confirm initial operation and power-up test cycle. Once there's confirmation of a power LED, wait a few minutes to confirm a status LED (if applicable) and the radio LED. Connection is typically made automatically in default mode to the other factory default radio.
4. If any one of the LEDs doesn't illuminate (sometimes it takes up to 2 minutes), then the unit is defective.

SYSTEM-WIDE TESTING As part of the system-wide testing for the Final Acceptance Test plan, a QA inspection checklist and/or report are required to guarantee that the following for each component was tested and verified:

1. Verified proper operation of the power supply.
2. Verified proper operation of the digital video encoder for video streaming and network connectivity.
3. All hardware (screws, nuts, bolts, lock-washers, etc.) is installed properly and tightened.
4. Cable ties and adhesive clips are secured and clipped.
5. Harness assemblies are neatly dressed, identified, and properly routed.
6. Power breakers and fuses are properly installed and are the proper value and size.
7. Top bracket assembly is properly mounted and secured to the upper case.
8. Terminal connections are properly identified and seated securely into the connector.
9. Surge suppressor is properly installed and plugged in securely.

In addition to the QA testing previously listed, conduct a system-wide test comprised of PING tests (which is a simple PING of the IP address of the camera or digital video encoder) that send live images across the network to command and control facilities (viewing video streaming through the VMS software), and control camera PTZ functions during and post installation. These system-wide tests were already performed before, during, and after the on site installation of the components, otherwise they wouldn't be live in the VMS software. The key here is to document those tests for a signature. If the implementation is smaller, then another testing procedure won't be an overwhelming task; for the larger projects, it could take days or weeks (usually not in the budget).

The system-wide camera tests should include a camera re-initialization test, which is rebooting the camera remotely (through the VMS software), verifying normal startup operation, and (if applicable) using the VMS software to operate the PTZ functions.

NETWORK DVS ARCHIVING TESTING Once all of the individual component tests are complete and it's verified that all cameras and control systems are operational, the next step is to confirm that the VMS software is recording the video streams. This next step tests the ability to retrieve recorded video. If the recorded video is there, from the moment the camera was added into the VMS system for archiving, then testing is complete.

Chapter Lessons

- Implementation of such a complex project as DVS is always a challenge, depending on the size.
- As with any project, the project manager's ability to manage the resources, whatever the size, is an important first step in a successful installation.
- It's important to understand that the project manager is the glue that holds the project together and that whatever support can be provided is going to reflect on everyone involved in the project.
- The people out in the field, working on electricity, networking, and installations and testing, may see the potential problems first hand, but it's the project manager who documents them in a status report and needs to explain any problems face to face with the customer. Depending on the customer, the project manager's explanations can make a good day or a very, very bad day.

10

Security Integration and Access Management

Many of the VMS software packages, including those bundled with the less expensive DVR card solutions, provide a means of integrating or adding messaging, alerts, or alarms to select events such as motion detection, tampering, or even access control. The more complete VMS suites, such as Genetec Omnicast, Milestone, OnSSI, or ipConfigure, have more extensive integration capabilities provided by the addition of a Software Development Kit (SDK). The SDK provides an application programming interface (API), scripts, code, and/or a development or configuration interface between their core software solution and third-party applications.

The more demands set on the VMS, the more sophisticated the integration, unless you choose VMS software that has the ability to provide those features or can seemingly integrate with select third-party applications (Figure 10-1).

Security Integration

The reason security systems, access control management, and video surveillance systems are converging into the IP network is that it's scalable and far more cost-effective. As mentioned in previous chapters, any implementation that requires dedicated connections between two devices (including CCTV) requires expensive cabling. It's not the cables that are expensive – coax and even fiber can be only cents per foot – it's the installation of those cables that's costly. The other reason for this welcomed convergence is the implementation of another proprietary system. Just as the single black-box digital video recorders used in video surveillance are unique to their manufacturers, so are access control systems. Any expansion requires the same equipment and possibly even the system integrator, depending on the growth of that system. If the system manufacturer is out of business or has discontinued the system, you're out of luck, so the convergence of all security systems into existing standards is a welcome change in the technology, implementation, and business process improvement.

IP-based systems are open platform, using Ethernet connectivity, which is literally a worldwide standard. IP networking provides more choices in hardware and software that use the existing infrastructure of the company's network. IP-based systems are future proof because they use interoperable (open platform) components that make it easier to take advantage of new advances in IP cameras and digital video encoders, access control systems, security systems, and general computing growth. As mentioned in

FIGURE 10-1 The Intelli-M Supervisor software interface incorporates video surveillance and access control using an IP-based solution.

Chapter 3, a DVR may be limited to the size of the hard drive inside and the processing power and memory embedded within the proprietary system; thus replacing it may cost thousands of dollars when an additional hard drive with a storage area network (SAN) only costs $100 and provides the ability to upgrade other components.

The existing networking infrastructures make it possible to replace legacy systems with IP devices (e.g., IP cameras and biometrics). IP networking also makes it possible to upgrade those legacy systems, such as plugging an analog camera into a digital video encoder, thus maximizing the original investment in hardware and infrastructure installations.

Video analytics can also be used throughout the integrated security system. For example, an employee swipes his ID card and exits the building. The access management system sends a message to the integrated video surveillance system to pan, tilt, and zoom (PTZ) the closest camera to that exit door, where another employee decides to tailgate and exit without swiping his security ID card. Video analytics software recognizes this event and sends a message, along with a video clip of the incident, to the centralized security office presenting it onto a prominent monitor for display. Security personnel follow up from there. Video analytics applications include door movement; determining location, speed, and direction of travel; identifying suspicious movement of people or packages; license plate identification and facial recognition and more as the technology expands. Video analytics provides the necessary intelligence to monitor for certain events, so the security personnel don't have to watch monitors 24/7.

Centralized Security Management and Monitoring

IP networking provides the infrastructure for centralized security management, not just a single security office where all the proprietary cables terminate into system controllers. Scalability isn't just for the expandability of the individual systems; it's also for the centralized security office. If a larger room is needed because everything is connected to the IP network, that security office can move anywhere there's connectivity without the need to rewire anything. Such a move isn't limited to the same building. If there's connectivity into the local area network (LAN), there's a connection to all the security systems (e.g., access control, video surveillance, fire, etc.) from anywhere. The IP network-centric security system can also be monitored by remote access, outside the LAN through a secure virtual private network (VPN) connection, so the physical buildings aren't considered a boundary.

These IP-based security systems only need an Internet connection to provide access through any device, such as the desktop computer or laptop at home, or a PDA and/or smart phone while on the road. Multi-tiered user account authorization levels allow granular control of access privileges into this system, just like the security hierarchy that exists within the network operating system.

High-end VMS systems such as Milestone Systems XProtect Enterprise, Genetec Omnicast, and OnSSI can integrate a large number of IP systems and manage them from a single interface. An advantage of centralized management for access and video surveillance systems is the accountability of a single server authentication location. User management provides real-time access information, location, and duration. If a PTZ camera is moved toward an exit, the centralized system provides the administrator not only with the user who has moved the camera's area of coverage, but also any information on all personnel using that specific exit.

Digital search and retrieval provide a faster, more accessible archive that can be searched sequentially by time or incident, enabling fast information retrieval and protection for intelligent response or evidence and accountability of users.

Integration Using I/O

As more security systems move toward using IP networks, the more open those applications will be to integrate into other IP systems (see Figures 10-2 and 10-3 for examples). The key is the interface between the applications. This interface could be hardware or software and as simple as a software driver or as complex as a proprietary hardware translator.

The newer VMS applications are built on the Windows platform, making it more accessible because of the wealth of the development community. This provides an

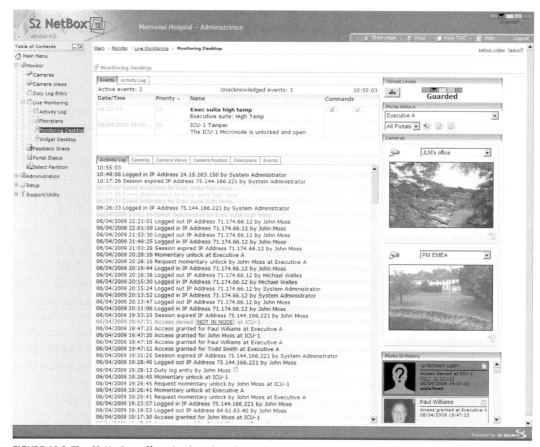

FIGURE 10-2 The S2 NetBox offers simple IP-based installation and integration of the access control system and OnSSI.

intuitive SDK that offers application developers and/or network engineers the ability to link the VMS software with other systems.

This software solution only works if there's a device in place functioning as an electronic relay switch or as a dry contact bridge. A digital video encoder includes communication terminal connections for these applications. In addition to servicing PTZ controls by using serial connections, digital video encoders also provide a power relay switch for communications.

Electronic Relay Connections

The relay switch connections are usually labeled COM, NC, and NO:

- COM: Common, always connect to this; it's the moving part of the switch.
- NC: Normally Closed; COM is connected to this when the relay coil is off.
- NO: Normally Open; COM is connected to this when the relay coil is on.

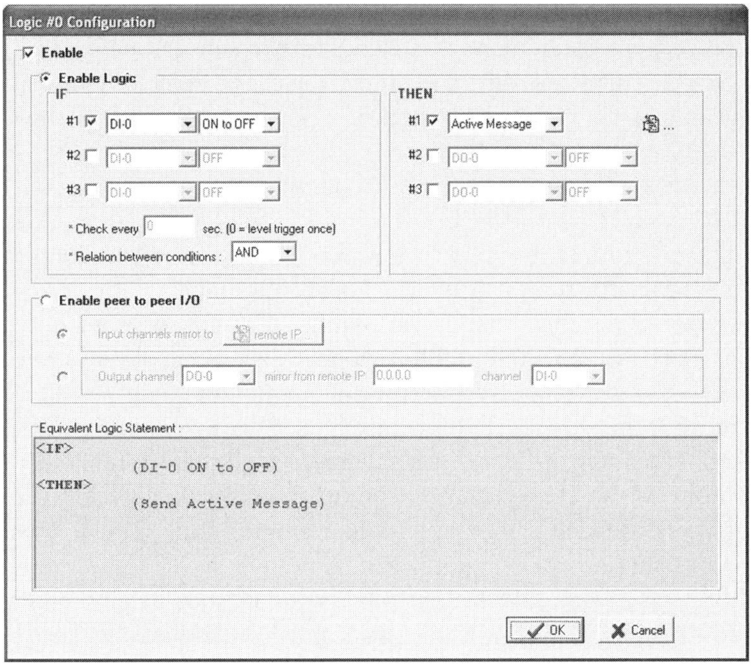

FIGURE 10-3 The Moxa I/O IP server provides an easy-to-use interface for the integration of security systems into the IP world.

Connect COM and NO together if you want the switched circuit to be on when the relay coil is on. Connect to COM and NC if you want the switched circuit to be on when the relay coil is off.

Since electronic relays are used as switches and follow the same terminology, they activate one or more poles when the contact is energized. This is done in one of three basic ways.

Normally Open

The relay switch is activated when the circuit is connected or the circuit closes. Otherwise, if disconnected, or normally opened (Figure 10-4), the relay stays inactive. This is also known as a Form A or "make" contact. A motion sensor is an example of how an NO relay is used. When there's no motion the circuit stays open, but if motion is detected, the motion sensor sends a signal to the relay, closes the contact, and activates the switch to turn on a security light.

Normally Closed

NC works in the opposite fashion as NO by closing the circuit or turning on the switch when the contacts are open and the contacts disconnect the circuit when the relay is activated or closed. It's also known as a Form B or "break" contact. An

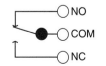

FIGURE 10-4 Relay circuit on NO.

example of an NC relay is the magnetic sensors used for windows and doors. The circuit is always closed because the door stays closed. The two sensors are pressed together and are normally closed. This is the dormant state. However, when someone opens the door and the circuit is opened, or disconnected, the power relay sends a signal to turn the switch on.

Change over (CO), or double throw (DT), contacts control two circuits: one NO contact and one NC contact with a common terminal (see Figure 10-5). It is also called a Form C or "transfer" contact (break before make). If this type of contact utilizes a make before break functionality, then it's called a Form D contact.

If the integrated system uses electromechanical relays, it includes one or more individual switches (each switch is referred to as a pole). Each one of these switches or poles can be connected or thrown together by energizing the relay's coil. This action is described as make (M) or break (B); for example, a simple relay with one set of contacts would be described as Single Pole Double Throw (break before make) or SPDT (B-M). These are examples of the more common contact types for relays in circuit or schematic diagrams, but there are many more possible configurations. The following are more common relay designations (see Figure 10-6):

Single Pole Single Throw (SPST) has two terminals that can be connected or disconnected. SPNO and SPNC are sometimes used to indicate if the connection is normally open or normally closed.

Single Pole Double Throw (SPDT) is a single common terminal connecting to one of two others.

Double Pole Single Throw (DPST) includes two pairs of terminals. They're equivalent to two SPST switches or relays actuated by a single coil.

The S or D may be replaced with a number, indicating multiple switches. For example, 4PDT indicates a four pole double throw relay.

General Alarm Connection

Alarm systems are run by a controller sometimes referred to as a switcher panel or board. These controllers are usually programmable logic controllers (PLC) defined by the National Electrical Manufacturers Association (NEMA) as a digitally operating

FIGURE 10-5 Terminal on back of digital video encoder.

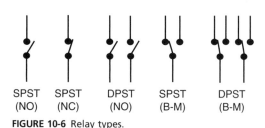

SPST (NO) SPST (NC) DPST (NO) SPST (B-M) DPST (B-M)

FIGURE 10-6 Relay types.

electronic apparatus that uses a programmable memory for the internal storage of instructions for implementing specific functions such as logic, sequencing, timing, counting, and arithmetic, to control various types of machines or processes through digital or analog input/output. The PLC accepts inputs from switches and sensors (measures or senses the system), evaluates these based on a program (logic), and changes the state of outputs to control a machine or process.

The controller may include a number of fused outputs for switching on or off security devices such as magnetic locks, door strikes, magnetic fire door holders, and/or digital video encoders to signal an event. As an example, let's say the controller has a choice of two alarm outputs. The controller may also have a choice of two inputs, NO or NC, and two dry contact outputs for a power fail/fuse blown alarm condition. The power supply inputs are either 12 VDC or 24 VDC, with the inputs protected by an end-of-line (EOL) resistor (if not, one should be added for termination; Figure 10-7).

When DC power fails or is disconnected, the dry contact outputs cause the relay to de-energize, thus sending a signal for an event trigger. NO and NC are used to describe how the relay contacts are connected and how to create the required actions. These input/output (I/O) interface modules or contacts are designed specifically to interface with computers, microcontrollers, and/or controllers to outside switches. There are four basic types of I/O modules available: AC or DC input voltage to transistor-to-transistor logic (TTL) or CMOS logic level output, and TTL or CMOS logic input to an AC or DC output voltage. These are available as individual solid-state modules or integrated into multiple channels.

FIGURE 10-7 Dry contact inputs on an alarm controller with EOL resistors.

There are three main types of I/O modules: the conventional rack-mounted type, which typically plugs into a backplane and requires a separate power supply; block I/O modules or I/O solid bricks, with a set number of inputs or outputs, or modular I/O units with a single communications controller; and stackable I/O add-on modules.

The conventional rack-mounted I/O modules typically also include the controllers and are better suited for mounting within a cabinet rather than an exposed rack. Block and modular I/O units are used inside or outside racks or cabinet configurations.

Flexible modular I/O units are plugged into an Ethernet network and controlled remotely.

Modular I/O allows for uniform, granular channel controls, plus scalability. When more connections are needed, another module can be added.

Alarms and Events

Events are triggered actions by inputs from either built-in devices such as digital video encoders, or third-party systems, such as point-of-sale terminals, alarms, or smart video software. The majority of the basic home or office alarm systems uses an I/O connection, so once the alarm is triggered – a magnetic door sensor activated, the fire alarm sensing smoke, or motion detected – the I/O input is recognized and a programmable event can be activated.

The key to integrating these systems is to analyze the details of how the system communicates and what, if anything, is needed to have the VMS software recognize the message – a pulse, code, or change in the current status.

Alarm and event management requires more sophisticated VMS software and although the more economical solutions provide motion detection, without I/O inputs on the hardware compatible with the VMS software it would be impossible to "read" the I/O contacts without some hardware interface. The key to successful alarm and event management is the ability to configure the VMS system to automatically respond to the event by sending alert notifications, recording video, displaying the designated video stream for that location with screen priority, or activating different devices such as doors and lights (Figure 10-8).

Another key aspect of DVS is the intelligent video functions that enable the VMS system to be more efficient, use network bandwidth and storage space better, and send alerts when an event occurs, rather than making the futile attempt to monitor numerous live cameras 24/7. All alarm and event configured responses can be automatically activated, thus helping operators cover more cameras and greater area of coverage.

FIGURE 10-8 Panic alarm pop-up message in Genetec Omnicast.

The more sophisticated VMS software can also manage the built-in event functionality within the DVS devices, or at least allow that device to properly run its functions without interference so that it can generate whatever responses are programmed from the device. In such a case, features like motion detection and/or camera tampering functions can be performed by the DVS hardware, which then sends the VMS software a flag for further actions, leaving this intelligence to the hardware and not the operators.

In those implementations where bandwidth is an issue, programming these features within the DVS system can reduce bandwidth and storage space by only streaming recorded and live footage during a triggered event. There's no need to continuously stream video to the VMS server for analysis at a later time, especially if 99% of the footage includes no motion at all. The video analytics happen at the device and not at the server. Also, processing these intelligent analytics is very processor-intensive, which not only reduces the life of the VMS server but its performance as well. With the digital video encoders doing the analytics, the VMS system can manage more cameras.

Alarm and event management includes digital video encoders and IP cameras constantly on guard analyzing inputs to detect an event and automatically responding with programmable actions provided by the VMS software. These events can be scheduled or triggered by the I/O ports on an IP camera or digital video encoder, which are connected to external devices such as a motion sensor, door switch, or fire alarm. Motion detection within the live video stream can also trigger an event like camera tampering, which is a relatively new feature that allows the camera to detect when it's been covered, moved, or is out of focus. Several more contemporary IP cameras and/or digital video encoders also include an audio trigger that triggers an event if sound is over or below a designated threshold as well as measure temperature by monitoring a programmed threshold (e.g., fire) before sending an event (Figure 10-9).

Event Responses

DVS camera and digital video encoders, with or without the assistance of VMS software, are configured to respond to event actions. When an event is triggered, responses can be scheduled or permanent. For example, if the VMS system is designed to be triggered by the activation of an emergency exit door, it can turn the closest PTZ cameras to that direction, both inside and outside the doorway; it can send a message that appears on the screen at the security office and if it's a selected door, an e-mail can be sent to management.

Another response, depending on the VMS software and devices used, includes an image or live or archived video footage uploaded to a specified location on the network. The I/O output ports on the IP camera or digital video encoder can be linked to external alarms, both audio – using sirens and/or digital audio files on the VMS workstation – or visual alarms such as specialty lighting. At Navy Pier in Chicago, when selected events are triggered, the lighting within the security office changes hue, each one color-coded for unique responses.

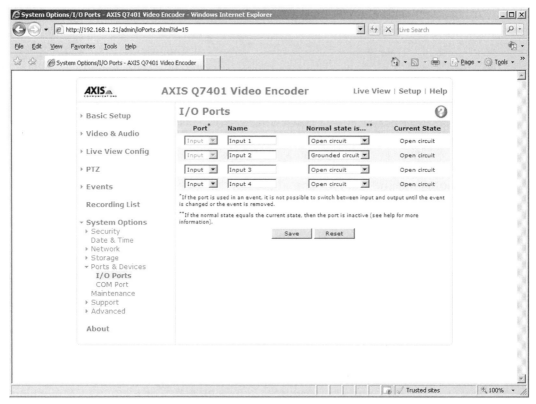

FIGURE 10-9 Setting event triggers using an Axis network video product Web interface.

Once an event is triggered, an e-mail is sent to notify personnel of the occurrence with a special notice and/or image capture attached. The difference between the e-mail notification and the message notices is that the message notices are limited to those personnel who are actively monitoring the VMS client software or, if not urgent, may be stored within the database for future reference.

The use of PTZ cameras, although limited in resolution, can be of great benefit to event responses by creating and then using select PTZ presets. These presets can be focused onto specific areas that include alarms, so when the alarm is triggered the camera pans, tilts, and zooms into that area of coverage. Depending on the size of the area of coverage, the camera may already be pointing in that location, which will then allow the viewer to see what occurred a few seconds prior by using the video archived footage. Figure 10-10 depicts the monitoring of the Smith Museum of Stained Glass Windows using a Genetec Omnicast interface at Navy Pier in Chicago. The museum exhibits have integrated sensors that trigger an alarm if the exhibit is touched. Those dry contact sensors are connected to several Moxa modular I/O servers that, upon activation, send a signal to the VMS software, which then triggers particular alarm responses. One response is to pan, tilt, and zoom the closest cameras to the activated

FIGURE 10-10 The Smith Museum of Stained Glass Windows is wired with sensors linked to the video surveillance cameras *(image courtesy of Navy Pier, Chicago).*

exhibit and display that video within an armed tile on the VMS interface, replacing the video on display at the time. The second armed tile displays the PTZ camera with the best proximity (and the one most likely to be watching that specific exhibit) and shows video footage from the archives for 3 seconds prior to the alarm. Most of the time an alarm is triggered by a tourist who has stepped too far back to take a photograph.

Figure 10-11 displays a small portion of the active functional diagram of the Museum of Stained Glass alarm triggers. Each cabinet is labeled with its original numbering scheme and the triangles represent each PTZ preset on each camera so that the VMS software recognizes which direction to point the camera in the event of activation of a shock sensor. Unfortunately, this depiction is black and white, while in actuality this diagram provides color-coded camera types, wiring colors, and two numeric codes per preset position: the actual preset number (within the VMS software) and the ASCII code sent by the Moxa server identifying which shock sensor in what display cabinet was activated.

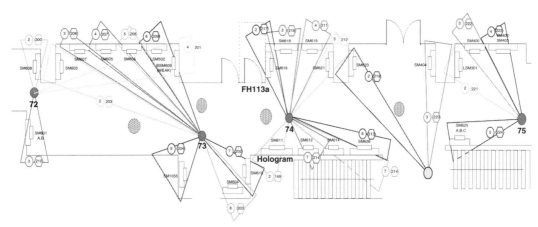

FIGURE 10-11 Section of the Museum of Stained Glass Windows functional diagram *(image courtesy of Navy Pier, Chicago).*

Whether using e-mail, an on-screen pop-up message, or sending a short message service (SMS) or multimedia messaging service (MMS) to a cellular device, each message includes specific procedures for the recipient to follow such as how to acknowledge the alarm.

INPUT/OUTPUT PORTS

The integrated I/O ports are unique to IP cameras and digital video encoders. As discussed in the Electronic Relay Connections section, the I/O ports enable network manageability of external devices through the VMS software. These are also used by configuring an IP camera or digital video encoder to record video only upon triggering of the external alarm sensor via its input port. This can save storage space, bandwidth, and research. A motion detection alarm may activate upon all movement, but the connection to the external alarm guarantees the footage is only the required video surveillance (Figure 10-12).

The basic rule for integrating external devices through the I/O ports of the IP camera and digital video encoder is if the device can switch between an open and closed circuit. The main function of the IP camera and/or digital video encoder's output port is to trigger external devices automatically, via remote control by personnel or by the VMS software. See Tables 10-1 and 10-2 for examples of devices connected to I/O ports.

Video Motion Detection

Video motion detection (VMD) is a common feature in many IP cameras, digital video encoders, and VMS software systems. VMD analyzes image data by differentiating from a series of images to determine motion in any part of a camera's view. VMS software also allows the user to configure inset windows within the video frame, such as a specific doorway, and/or excluded areas within larger inset windows. While many contemporary cameras provide this functionality, it isn't always completely compatible with the VMS software.

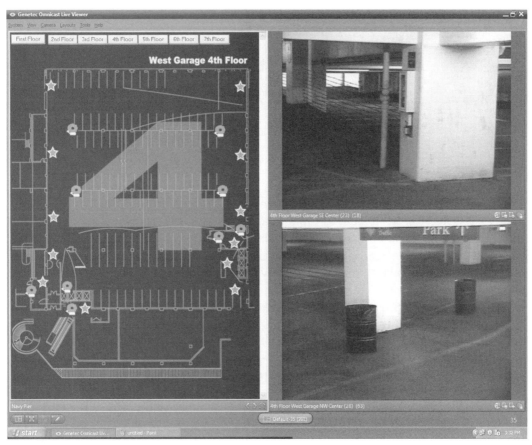

FIGURE 10-12 Parking garage fourth floor map with locations of each camera and panic call box *(image courtesy of Navy Pier, Chicago).*

Table 10-1   Examples of Devices Connected to the Input Port

Device Type	Description	Usage
Door contact	Simple magnetic switch that detects the opening of doors or windows	When the circuit is broken (door is opened)
Passive infrared detector (PIR)	A sensor that detects motion based on heat emission	When motion is detected, the PIR breaks the circuit
Inertia shock sensors	An active sensor that detects tampering and/or shock	When shock is detected on displays, glass, and/or passageways, the detector breaks the circuit

Active Tampering Alarm

IP camera and digital video encoders, especially using H.264, have vast amounts of computing power. This opens the door for new innovations such as active tampering intelligence, which helps prevent an age-old video surveillance problem: What happens

Table 10-2 Examples of Devices Connected to the Output Port

Device Type	Description	Usage
Door relay	A relay (solenoid) that controls the door-locking mechanism	The door-locking mechanism can be controlled manually through the network or be an automatic response to an alarm event
Siren	Alarm siren configured to activate when alarm is detected	Siren activation either by using the video motion detection or using the digital input
Alarm/intrusion system	An alarm security system that continuously monitors a normally closed or open alarm circuit	The alarm system serves as an extension to the VMS software, enhancing the alarm system with event triggers

when someone spray paints the camera lens view? Depending on the location of the camera, this is easily witnessed, as the vandal simply walks up to the camera and covers the view with spray paint. If the camera wasn't monitored and/or if the video footage wasn't archived, then there wouldn't be any evidence of the event. A VMS system with an active tampering alarm will immediately sense the abrupt change in visibility and send an alarm to the VMS software, where any number of triggered responses can be activated, with video pulled from the archive by any number of cameras within the area. Active tampering activates during accidental redirection, blocking, defocusing, or being spray painted, covered, or damaged.

Electronic Access Control and Management

Electronic access control (EAC) is technology used to provide and deny physical and/or virtual access into a physical and/or virtual space. That can be the building, the main distribution facility (MDF), or the executive suites. EAC includes technology as ubiquitous as the magnetic stripe card to the latest in biometrics. It provides technology for

- Controlling who can go where and when
- Controlling traffic in and out of areas
- People
- Vehicles
- Accountability

The benefit for access control management systems is the increased overall security of the physical (and virtual) domain. Visitor control and accountability are also important with each event and transaction recorded. Instantaneous alarm responses and reports are generated by the system for management review.

Types of EAC include:

- Discretionary
- Mandatory
- Role-based
- Rule-based

Discretionary access may include upper management's decision on who can read, write, or execute select files and services. This is generally used in MDF and/or sensitive file facilities.

Mandatory access allows a user to create new information but doesn't grant them administrative privileges to that information. They can create but can't determine who can access and modify that information. This type of access is widely used in military and financial institutions.

Role-based EAC is one of the most common group-based physical access control systems. Access is determined by the user's role in the organization: middle management has access to the second floor, upper management is allowed access to all floors, and everyone else is limited to the first floor.

Rule-based is the most common modern EAC, usually IP-based, and is rooted in a predetermined rules configuration. This type of EAC is simply the rules outlined to monitor and control who can go where and when.

The general access control process may involve an employee swiping a magnetic card through a card holder or presenting a smart card to a reader; either way, data are passed from the card to the reader. The data, typically a numerical code, are sent from the reader to a data controller unit or control panel. That controller validates the who, what, where, and when of those data based on data format, programmed reader location, time, and whether the card holder has permission to access that specific location at that time. Once the controller determines if the card holder, time, and place are valid, access is either granted or denied by data sent back to the reader. A transaction record is logged into the host system for future audits and review.

■ ■ ■ ▬▬▬▬▬▬▬▬▬▬▬▬▬▬▬▬▬▬▬▬▬▬▬▬▬▬▬▬▬▬▬▬▬▬▬

Access Cards

The magnetic stripe cards used in a variety of applications have an embedded numerical code hidden as an invisible magnetic bar code on the magnetic stripe. The swipe reader can determine who, what, and when is accessed via that encoded number.

A smart card contains a computer chip programmed for the same application. Smart cards are typically made of thick plastic and can be read wirelessly by the smart card reader from several inches away or even through clothing. However, if a magnetic stripe card becomes damaged, it may still function, but if the microchip inside the smart card is damaged, it's completely useless.

▬▬▬▬▬▬▬▬▬▬▬▬▬▬▬▬▬▬▬▬▬▬▬▬▬▬▬▬▬▬▬▬▬▬▬ ■ ■ ■

An access control system reduces risk by limiting potential losses from theft and protects intellectual properties, staff, and visitors.

EAC systems reduce the overall cost of security, but because it's a somewhat omnipresent system, it can better monitor physical locations at any time. There's no need for eliminating and/or re-keying locks and replacing keys, because changing the keys is as simple as denying all access to that individual's numerical code.

Although electronically monitoring the physical and virtual environment is an important part of all surveillance and security systems, the one element lacking is apparent in the limitations of a completely numerical system: there's no guarantee that the person you let into the office is the person who's supposed to be using that smart card. This is where integration with the DVS system becomes even more powerful. As discussed in Chapter 1, the primary reasons for video surveillance are deterrence, establishing a capable guardian, efficiency in security deployments, and detection. Integration of an access control system provides the VMS system with additional automated information to successfully achieve those goals and create a centralized location for security management.

The Access Control Market

The EAC management market is dominated by Lenel, Honeywell, Integral, and General Electric (GE). These are typically traditional point-to-point or RS485 serial multi-drop solutions requiring a separate run for power. These solutions include smart cards, biometrics, and video surveillance integration.

As depicted in Figure 10-13, the conventional access control process begins when the card holder presents the ID card to the designated reader. These unique data are passed between the card and the reader and then to the controller panel or PLC. They are validated within logical memory, which may include an ASCII code, date, and time

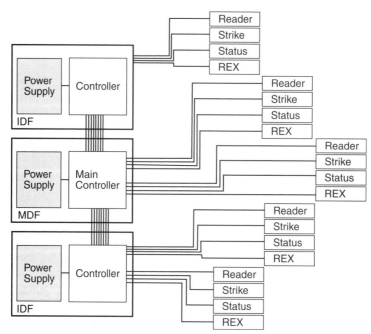

FIGURE 10-13 A traditional EAC topology.

and reader location, and then it's determined if that card holder has access to that location at that time.

The logic decision at the panel either grants or denies access and that decision is sent back to the reader where the confirmation LED either turns green for acceptance or red for denial. The transaction is then recorded to the host system and/or database where it can be saved as a report on disk or printed on paper.

The benefits of an EAC include a safe and secure environment with increased security awareness, visitor control, accountability, instant alarm responses, and reporting. There's also a reduction of risk from potential losses, theft of property or intellectual assets, personnel protection, and industrial espionage. The cost savings of an EAC system are gained from better security efficiency and from not having to re-key locks or replace keys.

The one key element of the EAC system that can't be overcome without either biometrics or integration into the video surveillance system is when someone steals or borrows another person's ID card. The analog world of CCTV made this proposition very costly, but with DVS it's more accessible. The most profound new EAC trend includes using an IP network-based infrastructure, which is similar to the paradigm experienced by CCTV. As a result, EAC can now also benefit from the significant reduction in infrastructure cost, modularity, and system expansion. An IP solution, much like an IP camera, only requires accessibility to an Ethernet network. This makes it even easier (maybe not simple, but easy) to integrate into the VMS, many of which now include EAC for future IP-based convergence.

These IP-based solutions include the necessary hardware for access management, typically using power-over-Ethernet (PoE), so that a single cable needs to be run to each end point. This Ethernet cable only needs to be terminated at the closest IDF location (much like an IP camera) to be plugged into the network. This eliminates the need to run cabling all the way back to a main controller.

The use of PoE for controllers, readers, and door releases requires backup power at each IDF location. Typically, these network or telecom closets, if equipped with mission-critical network equipment, already have UPS power sources. But power sources become even more crucial when physical access to the building, rather than access to the network, is at stake.

The IP-based EAC also provides connectivity to a single management tool interface, accessible from anywhere on the IP network. The use of the IP network also reduces installation costs and supports network discovery, DHCP, DNS, and static IP addresses.

The IP Network 802.1X IEEE standards use port-based network access control (NAC), which is included within the IEEE 802 (802.1) group of protocols and is increasingly deployed on corporate IP networks to provide closed wireless access for a select group of users. It's based on the Extensible Authentication Protocol RFC 2284 (recently succeeded by RFC 3748) and provides authentication to devices connected to a LAN. It's considered one of the best methods for securing wireless access points, denying access to all but the select few authorized to have access into the network. An IP-based EAC uses the same methods and hardware built into the network infrastructure (Figure 10-14).

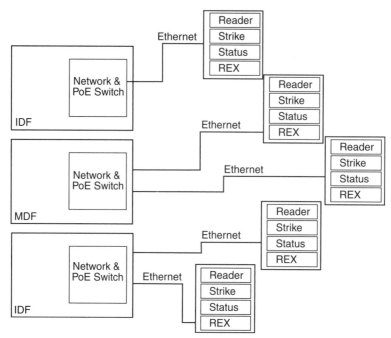

FIGURE 10-14 IP-based topology.

S2 NetBox is an IP-based EAC application that has also seamlessly integrated an IP-based access control solution into OnSSI and Milestone (Genetec has their own solution called Synergis). The advantage of S2 NetBox is scalability, beginning with a two-door system, which includes two secured smart card readers that can be linked into the IP network and powered over PoE. Figure 10-15 shows the monitoring screen available, once integrated with OnSSI. S2 NetBox Enterprise solution includes access control monitoring, video surveillance integration, temperature readings, locks, and a detailed activity log.

About Access Card Technologies

There are a number of door components used for access control of a physical location. The keypad, which is one of the more popular devices, uses a personal identification number (PIN) that allows or denies access. The keypad may also be part of a swipe (magnetic) or smart card proximity reader as a fail-safe method in the event the card becomes damaged.

The magnetic or smart access cards use an encoded electronic key that identifies the card holder. The magnetic access cards actually have a hidden bar code embedded within the magnetic strip. Although these are very common and low cost, they can be easily duplicated and/or damaged. The Wiegand magnetic access card uses ferromagnetic wires uniquely arranged to represent the encoded data. These cards make it difficult to duplicate, but they're typically special order items. The barium ferrite access cards use a

FIGURE 10-15 S2 NetBox Solution can use a map to monitor employee locations, room temperature, and even alarms.

laminated layer of magnetic material with dots to represent the encoded data. They cost less and are more durable than the magnetic stripe cards, but they also can be duplicated.

The proximity access card doesn't require physical contact (I have one and just keep it in my wallet and point it toward the reader). These are very difficult to duplicate because they use digital data (no analog magnetic stripe), but they're vulnerable to radio frequency (RF) interference and easy tailgating.

The smart card, either contact or contactless (requiring physical contact or just a proximity), stores a large amount of information on the embedded chip, and because of the available space it can be used for multiple purposes (e.g., access and parking). The smart card is the most versatile choice and is very difficult to duplicate, but it's also more expensive.

Biometrics

Biometrics is an unalterable personal characteristic or attribute like fingerprints, optical retina pattern, voice, or hand geometry. Although biometrics can eliminate the concern over access card sharing and card theft or loss, it still doesn't eliminate tailgating.

EAC System Topology

Figure 10-16 depicts the basic access control topology including the various components. The basic door configuration includes an entry request device such as a card reader, biometric device, or keypad. The door includes the locking mechanism controlled by the EAC management software. This door stays locked until access is granted. The door informs the management software when it was opened and recognizes when the door closes to re-engage the locking mechanism.

There may be a request-to-exit button or motion detector inside the facility to let the management software release the lock to exit. Typically these systems include an audible alarm that activates if the door is forced open or held open too long.

These data are saved and decisions are made by the logic inside a controller, which is independent of the management software. This logic component (see Figure 10-17) determines what to do when an event is triggered. The Moxa ioLogik I/O Servers can convert an event trigger result directly into a digital alarm output. This is easily set up using their Web-based ioAdmin interface to define an IF-THEN-ELSE logic rule, eliminating the need to write programs for PCs or controllers. The advantages of the Moxa I/O Servers are their independence from any VMS system and their IP-based communications, which makes it possible for integration into any VMS software that includes event and/or alarm management.

Integrated Access and Digital Video

The integration of digital video surveillance and access control is part of the natural convergence of security technologies. Centralizing security into a single interface offers additional information that has never have been possible before: notification of an employee, by name and identification number, entering the building; an image of that employee presented on-screen for verification (confirming it's not a stolen ID); and ensuring no tailgating occurs through image capture on the security camera.

The Genetec Security Center suite is a software solution incorporating all of these components into a single GUI. As depicted in Figure 10-18, the security center displays all of the security cameras as a tree in the left-hand pane, all employees currently in the building/office in the upper pane, and tiles of various video surveillance cameras. If an ID reader is used to gain access into select areas, the name and photographic image of that employee are presented in the lower right corner. This provides a simple way to confirm identity when the individual walks by the camera.

For these separate systems to converge or to integrate into a single interface, a detailed knowledge of all required systems is necessary. If there are no as-built documents providing functionality and configuration information to make this integration work, there are only two options. The first option, which may be the most complex, is to reverse engineer each system (if the original solution provider is unavailable or unwilling to provide those details) and document the topology and how it can be

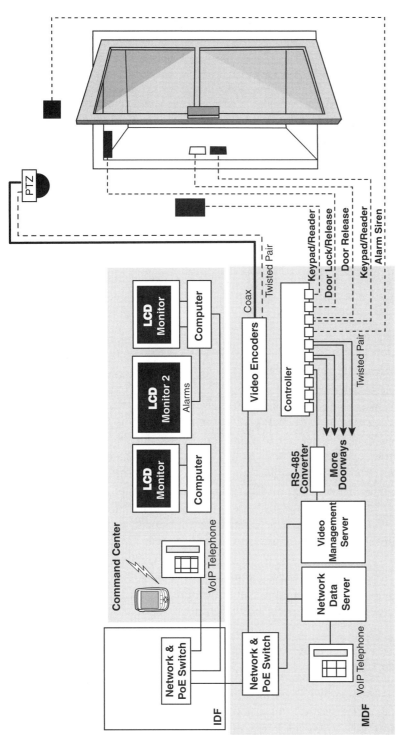

FIGURE 10-16 Traditional topology and components.

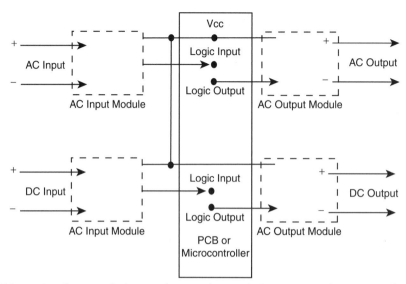

FIGURE 10-17 Somewhere between the input and output there's a logic component determining what to do when an event is triggered.

incorporated into the new DVS system. The other method is to rip and replace; a complete upgrade to a new security system with all the existing system requirements plus the ability to converge them all.

Figure 10-19 depicts the Genetec/IONODES Smart Vault, a unique network appliance that includes a Genetec software license for 32 cameras and up to 3 TB of storage. Targeted to small or medium businesses, this 32-camera solution provides BNC ports for up to 16 analog cameras (each new or existing), plus a built-in network switch for an additional 16 IP cameras, an IP-enabled access control system managed by the built-in access control software, and the ability to link multiple units together. Starting at an MSRP of $4800, which includes the software, this unit makes for a straightforward installation in a small, robust package. These network appliances provide a neat little package that brings the convergence of security assets to the smaller and mid-size companies.

These systems work together: video surveillance, access control, fire systems, elevator systems, intrusion detection, photo imaging, and building automation. Each system requires knowledge of architecture, operations, electrical, regulatory codes, and the customer's business needs (Figure 10-20).

Troubleshooting

It's difficult to troubleshoot digital I/O connections that are unique by manufacturer, design, and function. Because a DVS system involves the convergence of multiple devices under a single interface, there are many potential points of failure. Nevertheless,

FIGURE 10-18 Genetec Security Center.

the same three-dimensional (electrical, mechanical, and/or software) view is still applicable.

When integrating external systems into the VMS software, there's usually a trigger or multiple triggers that turn on the appropriate programmed function. For example, if motion is detected on the video image, the VMS software sends a trigger to the security lighting system and to the PTZ camera to shift the view to a select preset. The

FIGURE 10-19 Genetec smart vault.

security lighting illuminates the area of coverage and the PTZ camera follows the motion detected.

As with any troubleshooting, first ensure that all devices are powered. Most electronic equipment includes an LED light that offers an indicator of active power. If not, then a voltmeter would be required or spot-check the power for noise using an oscilloscope (especially if using a broadband-over-power lines solution). Don't ever take for granted that the wiring was installed and secured properly. I can't tell you how frustrating it is to troubleshoot a data connection only to find that the same installer tightened the same type of terminal onto the same wiring sleeve, neglecting to make contact with copper.

When troubleshooting physical wire connections:

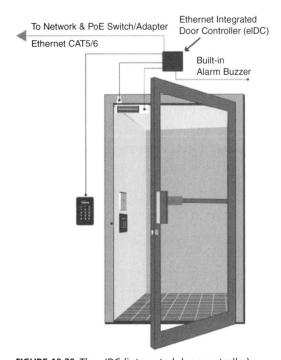

FIGURE 10-20 The eIDC (integrated door controller) uses a single Ethernet PoE cable for data and power distribution from any PoE network switch found in the facility.

- Check that connectors are screwed together tightly.
- Check to make sure the wiring is properly insulated.
- Check that outdoor gland connectors are secured.
- Check if foreign material is deterring copper connections.
- Check to make sure the wiring doesn't sit on heating devices.

Avoid twisting a bare wire and hooking it to a screw (unless using it as a ground). Don't twist the two ends together and then tape them as this creates future problems.

Once you've determined that the physical wiring is secured and properly insulated, the next step would be to determine if those wires are actually wired correctly for communications. You'll need wiring schematics for the two separate systems that require interconnectivity. For example, if the requirements include integrating the fire alarm system into the VMS software, then there need to be I/O ports on the fire alarm controller panel and either the VMS server or the digital video encoder. When the fire alarm activates, a trigger is sent to the VMS software to bring up the cameras in the location of the first alarm. The wiring schematics should provide details on how the two systems are configured and whether integration is even an option within an intermediary interface and/or translator.

Ohm Meter and Continuity Tester

An ohm meter is used to measure resistance and is very useful when troubleshooting wire circuits. An infinite meter reading on any test shows the circuit is open. All tests conducted with an ohm meter must be done with all power disconnected. There's also a continuity tester, which is a small inexpensive device whose only function is to determine if a circuit is open or closed.

If those wires are installed properly, the next step is the software interface responsible for managing the trigger and events. This involves verification of the software configuration, which may include online research and possibly a call to the device manufacturer (most user manuals aren't written with enough detail to be helpful for every conceived application).

Figure 10-21 provides a general path to determine if the problem may be mechanical, electrical, or software related. Every installation is different, and whether or not it's the actual event trigger, the network, or the two separate systems, there's always something that makes it a challenge each time.

Chapter Lessons

When it's time to converge security systems, here are a few key factors to remember:

- Integration of the security management systems increases employee efficiency and improves security.
- If upgrading the VMS and EAC systems together, research the various options that can satisfy system requirements and are already integrated and compatible.
- Choose the VMS software with the flexibility to accept and understand any external security system event triggers either directly or through an intermediate interface.
- There are many options available, with even more being developed, that can satisfy any system requirements no matter how large or small the implementation. There's no reason not to converge.

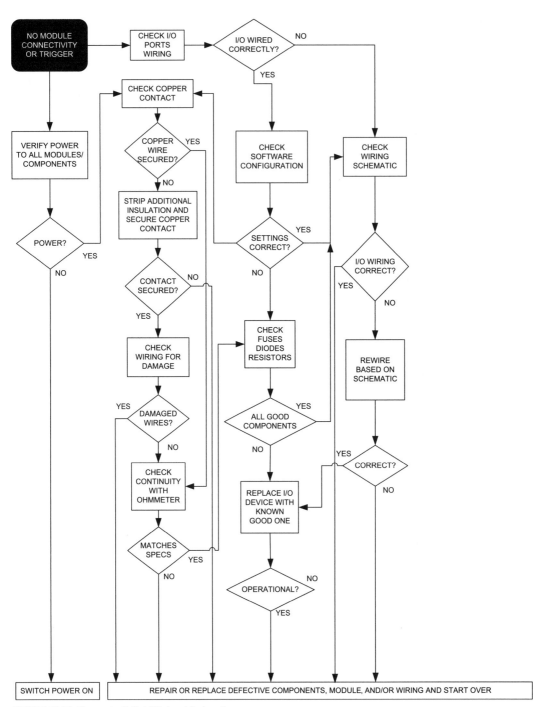

FIGURE 10-21 General: digital I/O troubleshooting.

Appendix: Site Survey Readiness Checklist and Survey Form

Presurvey (Readiness) Checklist (much of this information should already be in the statement of work)

Location: _____

_____

Address, City, State, Zip:
Can equipment be shipped to this location? Yes ___ No ___
Name of Recipient: _____
If no, what is the shipping address? Yes ___ No ___

_____

Address, City, State, Zip:
Location Contact Information

Managerial Name:		
Title:		
Phone:		
Mobile:		
Fax:		
E-mail:		
Available Hours: Janitorial/Maintenance Contact Name:		
Title:		
Phone:		
Mobile:		
Fax:		
E-mail:		
Available Hours:		

Background check required? Yes ___ No ___
Location Identification Card Required? Yes ___ No ___
Escort Required? Yes ___ No ___

General System Requirements:
Video Surveillance: Yes ___ No ___
Video Archiving: Yes ___ No ___
Video Analytics: Yes ___ No ___
Integrated Access Control: Yes ___ No ___
Decoder Requirements: Yes ___ No ___
Indoor ___ Outdoor ___ Both ___
Number of Existing Cameras:

ID	Type	Quantity
1	Analog PTZ	
2	Analog Fixed	
3	IP PTZ	
4	IP Fixed	

Interior Location of Existing Cameras (e.g., egress, elevators, hallways):

ID	Location	Quantity

Number of Existing Outdoor Cameras (e.g., parking lot, common areas):

ID	Location	Quantity

What's the required naming convention for cameras?

_____

_____

What are the lighting conditions requirements?
___ Day
___ Low-Light Night
___ Pitch-Black Night
___ Day and Night ___ Low-Light ___ Pitch-Black

What is the video resolution requirement?
___ CIF
___ 4CIF
___ 1.3 MP
___ 3 MP
___ Other _____
___ Don't Know

What's the compression requirement, if any?
___ MPEG-4
___ MJPEG
___ H.264 (MPEG-4, v10)
___ Don't Know

What are the frames per second (fps) requirements?
___ 5-7 fps
___ 10 fps
___ 15 fps
___ 30 fps
___ Don't Know

What is the video retention period?
___ 1 Day
___ 7 Days
___ 14 Days
___ 30 Days
___ Other _____
___ Don't Know

Provide a detailed summary for the archiving storage space availability:

Detailed power requirements (electrical, grounding, and power systems):

Mechanical and HVAC systems:

Environmental, fire, and security protection:

Camera Site Survey Form (Duplicate per Camera)

Camera Number:	
Logical Name:	
Description:	
Camera Type:	___ PTZ ___ Fixed
Camera Housing:	___ Indoor ___ Outdoor ___ Special
Mounting Bracket:	___ In Ceiling ___ Corner ___ Parapet ___ Roof
Power Requirements:	___ PoE ___ 24 V ___ 120 V ___ Other
Vandal Resistant?	___ No ___ Yes
Wireless?	___ No ___ Yes
General Location:	
Area of Coverage: (Photo of each object/location)	
NOTES:	Photo 1

Continued

Camera Number:	
NOTES:	Photo 2
NOTES:	Photo 3
NOTES:	Photo 4
Required PTZ Camera Presets?	

Location on Map (if applicable):

Please provide a summary of the lighting conditions:

Video Transmission Methods

Coaxial Cabling Requirements:

Cable Length	Distance to IDF	IDF	Encoder	New of Existing Conduit

Fiber Optics Cabling Requirements:

Single or Multimode	Transceivers	Distance to IDF	IDF	Fiber Termination Port	New or Existing Conduit

Twisted Pair Cabling Requirements:

Cable Length	Distance to IDF	IDF	Category	New or Existing Conduit

Mounting/Location Details (Photo)
Cabling Requirements (Wiring Diagram)
Wireless Information:

Proposed Wireless Networking Technology:	
Radios used for site survey:	
Height of survey antennas:	
Spectrum analysis: (note any possible interference)	
North:	
East:	
South:	
West:	
NOTES:	

Radio Number	SNR and/ or RSSI	Antenna Type	Antenna Height	Antenna Orientation	Coaxial Cable Length	Polarization
Radio 1						
Radio 2						
Radio 3						
Radio 4						

Radio Number	Frequency	SNR and/or RSSI	Frequency	SNR and/or RSSI	Frequency	SNR and/or RSSI
Radio 1						
Radio 2						
Radio 3						
Radio 4						

Index

Note: Page numbers followed by *f* indicates figure, *t* indicates table and *b* indicates box.

A

Abandoned object detection, 207
Access cards, 295*b*, 298, 299
Access point (AP), 123–124, 126–131, 148, 187, 188
  radios, configuring, 150–151
  topology, wireless, 127*f*
  wireless, 148
Active systems, 1–2
Adapters, 73*b*, 73*f*, 94
ADC. *See* Analog-to-digital converter
Address resolution protocol (ARP), 91, 92
Adobe Photoshop, 29, 30, 30*f*, 192
Advanced Systems Format, 31–32
Alarm(s), 286–292
  controller, 286–287, 287*f*
  linked to I/O, 289
  management, 288, 289
  notifications to mobile device, 292
  panic, 207, 210, 288*f*, 293*f*
  systems v. DVS, 11
  tampering, 206, 293–294
  triggers, 291, 292
Alternating current (AC), 124
Analog cameras
  digital v. megapixel v., 48–50
  IP camera v. digital video encoder and, 36–37
  PTZ, 58–59
  serial baud rate of, 81–84
  troubleshooting, 78–80, 80*f*, 82*f*
Analog video
  digital video v., 19, 20–21, 20*f*
  interlaced lines of, 23
  pixels of, 25
Analog-to-digital converter (ADC), 144
Antenna(s), 131–138, 139*t*. *See also* Specific types

cable, weatherizing, 159
coaxial cables for, 155–160
coaxial connector for wireless, 155, 155*f*
dBi of, 134–135
diversity, 144
EIRP of, 132
height of, 156
MIMO, 144
polarization of, 136
power output of, 135*t*
radio connected to, 134–135, 134*f*
for spectrum analysis, 188
systems, smart, 144
Anycast, 112
Aperture iris, 40*f*
API. *See* Application programming interface
Application programming interface (API), 281
Applications. *See* Software
Archiving, 34, 203, 211, 219, 220–221, 280
Area obstruction detection, 205
Area of coverage, 171–172, 182*f*
ARP. *See* Address resolution protocol
Artifacting, 162
As-built documents, 272–275
Aspect ratio, 27*t*
Audio, two-way, 63
Audio video interleave (AVI), 31–32
Autofocus, 45
Autotracking, 203, 204
AVI. *See* Audio video interleave
Azimuth pattern, 135, 135*f*

B

B. *See* Break
Baluns, 73*b*, 73*f*
Bandwidth, 107, 289
  digital video v., 108

Bandwidth (*Continued*)
 802.11n for high, 143, 144
 ethernet, 93, 97
 of fiber optics, 100
 of MPEG-4 v. MJPEG, 33
 pixels v., 31
 RF sharing of, 123
 streaming video v., 107
 unicasting v., 111
Basic service set (BSS), 127
Beacons, 128
B-frames. *See* Bi-directional predicted frames
Bi-directional predicted frames (B-frames),
  35, 37*f*
Binary, 19, 106*b*
Binary phase shift keying (BPSK), 143
Biometrics, 299
Bits, 25–26, 26*t*, 35, 104, 104*f*, 105*t*
Bitstream, 35
Boots, 75*f*
BPL. *See* Broadband over Power Lines
BPSK. *See* Binary phase shift keying
Break (B), 286
Broadband over Power Lines (BPL), 101–102,
  101*f*
Broadband telecommunication technologies,
  7, 8
Broadcast packets, 113–115
Broadcast storms, 113, 114
BSS. *See* Basic service set
Business process improvement, 197–199

C
Cable(s), 281. *See also* Coaxial cable; Digital
   video cables
 analog, 71–72
 crossover, 94, 94*f*, 99*t*
 ethernet, 74–77, 74*b*, 81, 94, 98, 99*t*
 fiber optic, 100*f*
 LMR195/RG 58, 156, 156*f*, 157*f*
  stripping tool, 157*f*
 network, 98–102
 stripper, 158
 tagging, 184*b*
 termination, 74–77, 74*b*

 twisted-pair copper, 98*t*
 weatherizing, 159
CAM. *See* Content addressable memory
Camera(s)
 alignment detection, 206
 analog v. digital v. megapixel, 48–50
 CCD v. CMOS image sensor of, 43
 choosing appropriate, 45–63
 circuit breakers for, 177–178, 177*f*
 day, 54–55, 54*f*
 day/night shots of, 55*f*, 56
 digital encoder compatibility with, 60–61,
  199–201
 dirt v., 56
 distance from, 174, 174*f*
 dome, 56
 drop down menu for, 209*f*
 electronic, 40
 fixed, 45
 fps of IP, 49
 fps of PTZ, 49
 LED light of, 88
 location of, 171, 181–183, 274
 long-range, 48
 megapixel, 48–50, 49*f*, 59, 172, 183
 night, 54–55, 54*f*
 NTSC IP, 48–49
 outdoor, 172
 outdoor/indoor, 58
 power, troubleshooting, 87*f*
 powering, 51–52, 177–179
 presets of PTZ, 70–71
 PTZ v. fixed, 58–59
 tampering alarm, 206, 293–294
 testing, 277
 thermal, 48
 video v. still, 42
 weather v., 56–58
 wireless everything, 149*b*
Camera/video site survey, 180–183
Capable guardian, 2–3
Carrier sense multiple access/collision
   avoidance (CSMA/CA), 140
CCD. *See* Charged couple device
CCIR 601, 31–32

CCK. *See* Complementary code keying

CCTV. *See* Closed circuit television

CDRL. *See* Contract Deliverables Requirement List

Central processing unit (CPU), 9, 226

Change over (CO), 286

Change request, 271

Channel planning, 150

Charged couple device (CCD), 40, 43, 172

CIF. *See* Common Intermediate Format

Class identifiers, 106*t*

Client relations, 180

Client/server networks, 94–95, 223

Climate control, 57*b*, 58

Closed circuit television (CCTV), 1–2, 3–6, 184*f*, 199, 209

  DVS conversion topology of, 10*f*, 184*f*

  installation, 5*f*

  interface, 199*f*

  upgrading, 60

Closing process

  camera testing, 277

  digital video encoder testing, 278

  final acceptance test plan, 277–280

  modular testing, 277

  network DVS archiving testing, 280

  network switch testing, 278–279

  system-wide testing, 279

  wireless radio testing, 279

CMOS. *See* Complementary metal oxide semiconductor

CMYK. *See* Cyan, magenta, yellow, and black

CO. *See* Change over

Coaxial cable, 71, 76

  antenna, 155–160

  signal loss of, 77, 77*f*

  termination of, 74–77, 76*b*, 76*f*, 156–160

Codecs, 31–32

Cohen, Lawrence, 2–3

Color. *See also* Digital color depth

  32-bit, 26

  depth of analog television, 21

  metadata, non, 26

  RGB v. CMYK, 21

Common (COM), 285

Common Intermediate Format (CIF), 31, 34, 48–49

Communications, 61, 62*b*, 108, 193

  fiber optics and, 100

  node, 129

  OS, 121

  point-to-point, 109–111

  synchronous, 6

Compatibility, 199–201

Complementary code keying (CCK), 140

Complementary metal oxide semiconductor (CMOS), 40, 43, 173, 287

Compression, 31–32, 108

  DCT, 35, 36

  H.264, 33, 37*t*, 50

  JPEG, 32

  MJPEG, 32–33

  MPEG, 34–36

Computer(s), 6–7, 91, 222–223, 224, 229. *See also* Networks

  adding a DVR card to, 229–232, 230*f*

  analog cables for, 71–72

  case, high-performance, 233*f*

  crossover cable for connecting, 94, 94*f*

  evolution of, 148

  farm, 97–98, 121–122

  power supply, 232

  processing power, 197*b*

  RTP communication between IP camera and host, 115*f*

  VM v. host, 116

  workstation v. server, 222–223

Conduit, exposed, 12*f*

Connectors, 71–77, 72*f*, 158*f*

  90 degree angled, 74, 76, 76*f*

  adapters and baluns for, 73*b*, 73*f*

  BNC, 71, 76, 162–163

  crimp-on, 158–159

  DVI-D, 74

  EZ, 74–77, 156, 157*f*

  fiber optic, 100

  F-Pin, 71

  HDMI, 74

  N-type, 156*f*

  RCA composite, 71

Connectors (*Continued*)
  RJ45, 75, 98*f*
  SMA, 156*f*, 157*f*
  troubleshooting, 79
  wireless antenna coaxial, 155, 155*f*
Content addressable memory (CAM), 96–97
Continuity tester, 305*b*
Contract Deliverables Requirement List
      (CDRL), 267*b*, 267*t*
Conventional rack-mounted I/O units, 287
Convergence, 7–8, 8*f*
Cooling system, 57
CPU. *See* Central processing unit
Crime, 1–2, 3
Critical Path Method (CPM), 261–262, 262*f*, 262*t*
Crossfire, 226–227
Crowds, DVS for large, 7
CSMA/CA. *See* Carrier sense multiple access/
      collision avoidance
Cyan, magenta, yellow, and black (CMYK), 21

D
Data
  flowchart, 161–162, 161*f*
  pathways, 124, 183–184
  power v., 178*b*
  real-time retrieval of, 204
  recovery, 247*b*
Databank, 203
Database, 201, 203
dBi, 134–135
dBm, 134
DBPSK. *See* Differential binary phase shift keying
DCT. *See* Discrete cosine transform
Decibels (dB), 133, 134–135
Decimal, to binary conversion, 106*b*
Depth of Field, 41–42, 41*f*, 45
Detection, 3
Deterrence, 1–2, 3
DHCP, 66–67
Differential binary phase shift keying
      (DBPSK), 138–139
Digital color depth, 25–26, 26*b*, 26*t*
Digital image sensors, 43, 43*f*, 44*t*
Digital signal processing (DSP), 144

Digital video, 19–20, 39
  analog video and, 19, 20–21, 20*f*
  appliance, 195–197, 197*f*
  archiving, 34
  bandwidth v., 108
  flickering of monitors v., 23
  formats, 31–33, 31*t*
  integrated access and, 300–302
  multiple streams of, 34
  networking, 108–113
  resolution, 26–27, 27*t*, 28*f*, 31
Digital video cables, 71–77, 72*f*
  adapter and baluns for, 73*b*, 73*f*
  analog, 71–72
  DVI-I, 74
  factory-made, 74
  single 8-pin Ethernet, 81
  stripping of, 74, 75*f*
  S-video, 71–72
  termination of, 74–77
  troubleshooting, 79
  VGA, 71–72
Digital video encoder(s), 23, 27–31, 62, 63*f*,
      68–70, 115, 119–121, 199–200, 286*f*
  2CIF, 31
  4CIF high-resolution analog video security
      camera with, 29*f*
  built-in, 45
  camera compatibility with, 60–61, 199–201
  configuring, 63–71
  default password of, 68
  default settings v. configuration of, 67
  DHCP enabled, 66
  drivers for, 121
  drop down menu for, 209*f*
  inputs of, 62
  installation and configuration applications
      for, 63–65
  IP camera v. Analog camera and, 36–37
  PING tool for testing connectivity of, 78–79
  PTZ camera connected to, 58–60, 81
  ruggedized, 62
  serial web interface of, 61–62, 61*f*
  software, 63
  testing, 278

troubleshooting an analog camera to, 78–80, 80*f*

Verint SConfigurator software for, 65–66, 65*f*, 66*f*, 121

VMS v., 78

wiring variations of, 61–62

Digital video interface-digital (DVI-D), 74

Digital video interface-integrated (DVI-I), 74

Digital video recorder (DVR), 195–196, 219

   card, 228–232, 228*f*, 230*f*, 232*f*

   card, troubleshooting, 242–249, 243*f*

   card compatibility with VMS, 232

   card v. fps, 229

Digital video security (DVS), 7, 9–10, 39, 202–203, 209, 234–235

   alarm systems v., 11

   analog PTZ camera in, 59–60

   applications, 105–106

   archiving, 219

   camera site survey process of, 191*f*

   convergence in, 7–8, 8*f*

   conversion to, 10*f*, 197–199, 198*f*

   design, 272

   Ethernet configuration for, 67–68

   Ethernet hub for, 97

   hardware, upgrading, 224–237

   beyond human capability at 30fps, 50–51

   incompatibilities within, 119

   intelligent video functions, 288

   multicasting for, 112–113

   network, isolation of, 185*b*

   network access of, 239–241

   poles of mounting, 12

   port numbers for, 110*t*

   powering, 9, 12*f*

   remote reboot of, 52

   remote viewing, 240–241

   RF v., 9

   site surveys, 171

   storage, 50, 220–222

   troubleshooting of, 52, 77–88, 80*f*, 116, 161

   unit discovery of, 63

   VMS requirements, 219–222

   wireless connectivity of, 164–167, 166*f*

Dipole antennas, 136

Direct sequence spread spectrum (DSSS), 129, 138–140

Directional antennas, 130, 132*f*

Discrete cosine transform (DCT), 35, 36

Disk management console, 247, 248*f*

   disk missing, 248

   disk not initialized, 249

   disk unreadable, 248

   errors of, 247–248

   foreign disk, 248

Display resolution, 29

Distortion, 23, 126

DNS. *See* Domain name server

Documentation. *See* Project documentation deliverables

Domain name server (DNS), 92, 104–105

Doorways, 172

Dots per inch (dpi), 25, 29

Double pole single throw (DPST), 286

Double throw (DT), 286

Dpi. *See* Dots per inch

DPST. *See* Double pole single throw

Drivers, 121, 199–200, 227*b*, 242

DSSS. *See* Direct sequence spread spectrum

DT. *See* Double throw

Due diligence, 124

DVI-D. *See* Digital video interface-digital

DVI-I. *See* Digital video interface-integrated

DVR. *See* Digital video recorder

DVS. *See* Digital video security

Dynamic host configuration protocol (DHCP), 148, 149, 223, 239

Dynamic IP address, 66

E

EAC. *See* Electronic access control

E-Box, 63

802.11. *See* Wireless local area network

EIRP. *See* Equivalent isotropically radiated power

Electronic access control (EAC), 294–302

   cards, 295*b*, 297, 298–299

   discretionary, 295

   integrated access and digital video, 300–302

   mandatory, 295

Electronic access control (EAC) (*Continued*)
  role-based, 295
  rule-based, 295
  S2 Netbox Solution, 298, 299*f*
  topology, 296*f*, 300, 301*f*
Electronic relay connections, 284–286, 286*f*
  COM, 285
  NC, 285–286
  NO, 285, 285*f*
Electronics, 57
Enclosures
  hardened, 57
  laboratory testing of camera, 57
  outdoor, 58, 63, 64*f*, 88
  residue on, 56
  vandal-resistant, 47, 48*f*, 56
Encryption, 154
  ESSID, 154
  key, 149, 167
  wireless security, 149*t*
  WPA/WEP, 148
End-of-line resistor (EOL), 287, 287*f*
Equivalent isotropically radiated power
      (EIRP), 132
Error variable, 36
ESS. *See* Extended Service Set
ESSID. *See* Extended Service Set ID
Ethernet, 45, 50, 60, 89, 92, 93–102
  bandwidth, 93, 97
  cable, 74–77, 74*b*, 81, 94, 98, 99*t*
  configuration, 67
  connectivity, 281–282
  familiarity with, 93
  network, 90, 119
  port, RJ45, 63
  star and tree topology for, 94, 103–106
  topologies, 93, 93*t*
Ethernet equipment, 95–98
  firewall, 97–98
  hub, 97
  router, 96
  switch, 96–97
Events
  alarms and, 288–289
  management, 288, 289

  responses, 289–292
  triggers, 290, 290*f*, 291, 302*f*
Execution process, 269–271
  change control process, 271
  escalation in, 270*f*
  project tracking, 269–270
  status meetings and status reports, 270–271
Expansion cards, 228–232
Extended Service Set (ESS), 127
Extended Service Set ID (ESSID), 154

F
Facial detection, 206
Facial recognition. *See* Human recognition
Fall detection, 206
FCC. *See* Federal Communications
      Commission
Federal Communications Commission (FCC),
      20–21, 125
FHSS. *See* Frequency hopping spread
      spectrum
Fiber optic(s), 11, 100–101, 100*t*
  bandwidth, 100
  bundles, 100
  cables, 100, 100*f*
  carrier frequencies of, 100–101
  RF v., 100–101
Filter, 30
Final acceptance test plan, 277–280
Firetide mesh configuration, 151, 152–153, 152*f*
Firewall, 96, 97–98, 148, 239–240
Firmware, 71, 85*b*, 200
5 GHz band, 141, 142*t*
Fixed cameras, 45, 58–59
Flat panel antennas, 137, 138*f*
Flexible modular I/O units, 288
Focal length, 42
Focus, 45
Form D contact, 286
Forward predicted frames (P-frames), 35, 37*f*
4CIF. *See* Four times Common Intermediate
      Format
Four times Common Intermediate Format
      (4CIF), 31, 48–49
F-Pin connectors, 71

fps. *See* Frames per second
Frames, 35, 36–37
Frames per second (fps), 22*b*, 22*t*, 42, 50–51,
    51*f*, 52*f*
  CIF, 31
  DVR cards and, 229
  GoP, 36
  IP cameras, 49
  worldwide standards for, 22
Frequency hopping spread spectrum (FHSS),
    129, 138–140
Fresnel Zone, 130–131, 131*f*
F-Stop, 40–42, 44
Full screen display, 29

G
Gantt chart, 252–253, 261–262
Genetec Security Center, 196, 212*t*, 300,
    302, 303*f*
GHz channels, ISM, 141*f*
Google Earth, 181, 187, 192, 192*f*
GoP. *See* Group of pictures
Grounding, 185–187
Group of pictures (GoP), 36

H
H.264, 33, 37*t*, 50
Hackers, 97–98, 121–122, 138–139, 240
Hand-holes, 12, 12*f*
Hard drive(s), 220, 232–235, 241
  adding additional, 233–235, 235*f*
  controller cards, 228
  cooling fans, 241, 242*f*
  maintenance, 241
  master and slave IDE, 234, 234*f*
  RAID, 235–236
  troubleshooting, 231*f*, 245–249, 246*f*
Hardened enclosures, 57
Hardware, 200
  incompatibilities, 119
  software v., 26*b*, 39
  system requirements, VMS, 224–225, 224*t*
  tools, 190, 193
HDMI. *See* High definition multimedia
    interface

HDTV. *See* High definition television
Heating system, 57, 58
Hertz, 125*b*
Hidden node, 128–129
High definition multimedia interface
    (HDMI), 74
High definition television (HDTV), 8, 19–20,
    25, 28*f*
Host(s)
  identifier, 104–105, 106
  number, 103
  unicasting to unique, 111*f*
HTML. *See* Hypertext markup language
HTTP. *See* Hypertext transfer protocol
Hubs, 97, 118, 222–223
  speed, 97
  of star network, 103
  T, 97
Human recognition, 173–176, 206
  database for, 203
  facial pixels v. resolution size in, 176*t*
  pixels of, 174, 175*f*, 176*f*, 176*t*
  resolution of, 174
HyperTerminal, 59
Hypertext markup language (HTML), 115
Hypertext transfer protocol (HTTP), 115,
    119–121

I
IANA. *See* Internet Assigned Numbers
    Authority
IBSS. *See* Independent Basic Service Set
IDE slots, 233*f*
IDF. *See* Interim distribution facility
IEC. *See* International Electro-technical
    Commission
I-frames. *See* Intra-frames
IGMP. *See* Internet Group Management
    protocol
Illumination, 53–56, 53*f*
Illuminators, 56
Images, 43, 275–276, 275*f*
Independent Basic Service Set (IBSS), 126, 127
Industrial, scientific, and medical bands
    (ISM), 140, 141*f*

Infrared (IR). *See also* specific types
  cameras, 48
  illuminators, 56
  technology, 54–56
Initiating, 251–252
Input/output (I/O), 284*f*, 287, 288, 293*t*
  alarms linked to, 289
  conventional rack-mounted, 288
  event trigger v., 302*f*
  flexible modular, 288
  modular, 288
  ports, 292
  troubleshooting, 302–305, 306*f*
Installation, 63–65, 180
Integrated access, digital video and,
    300–302
Integrated door controller (eIDC), 304*f*
Integrated drive electronics (IDE), 228
Integration, 283–284, 284*f*, 285*f*
Intelligent scene verification, 205
Intentional radiator, 132
Interference, 19, 124, 129–130, 136, 143, 188
Interim distribution facility (IDF), 5–6,
    183–185, 209, 222
Interlaced lines, 23, 23*f*, 24*f*
International Electro-technical Commission
    (IEC), 32
International Organization for
    Standardization (ISO), 32
International Standard Organization (ISO), 91
International Telecommunications Union
    (ITU), 31, 32
Internet, 19–20, 104, 106
Internet Assigned Numbers Authority
    (IANA), 112
Internet control message protocol (ICMP), 91
Internet Group Management protocol
    (IGMP), 113, 114
Internet protocol (IP), 89
Intra-frames (I-frames), 35, 37*f*
I/O. *See* Input/output
IP address, 66, 67–70, 103–106, 104*f*, 208
  ARP, 91, 92
  bits of, 104, 104*f*, 105*t*
  Class A network, 105

Class B network, 105–106
Class C network, 105–106
class identifiers, 106*t*
classes, 105–106
decimal numbers of, 92
duplicate, 67–68
format, 104
host identifier of, 104–105
multicasting Class D, 112
network identifier of, 104
number of, 106
static, 149
for TCP/IP, assigning, 103
IP cameras, 45, 50, 68–70, 81, 114, 119–121,
    199–200, 239, 292
  analog camera and digital video encoder v.,
    36–37
  compatibility issues of, 60
  configuring, 63–71, 69*f*
  drivers for, 121
  PTZ, 50, 54
  RTP communication between host
    computer and, 115*f*
  troubleshooting, 81, 82*f*, 83*f*
  VMS software v., 50
  web interface of, 70*f*
  wireless, 149
IP hosts, DHCP configuration parameters
    for, 66
IP networking, 89, 119, 120*f*, 209, 281–282,
    283, 297
IP version 4 (IPv4), 106
IP version 6 (IPv6), 106
IP Video System Design Tool, 174, 175*f*
IPCONFIG/ALL, 92, 92*f*, 117
IPv4. *See* IP version 4
IPv6. *See* IP version 6
IR. *See* Infrared
ISM. *See* Industrial, scientific, and medical
    bands
ISO. *See* International Organization for
    Standardization; International Standard
    Organization
ITU. *See* International Telecommunications
    Union

J

Joint Photographic Experts Group
 (JPEG), 32
JPEG. *See* Joint Photographic Experts Group

L

LAN. *See* Local area network
Laptop, 67, 116, 163, 190–193
LCD. *See* Monitors
Lens calculator, 174, 175*f*
Lenses, 39–40, 41
 climate v., 58
 fixed, 44
 focal length of, 42
 plastic, 46–47, 47*f*
 quality of, 46–47
 sensor sizes of normal, 44*t*
 spray painting, 293–294
 telephoto, 42
 of thermal cameras, 48
 wide-angle, 42
 zoom, 44–45, 44*f*, 47, 54, 56
License plate recognition (LPR), 172, 173,
 173*f*, 176*t*
Light, 39–40
 environments, low, 42
 IR, 55
 sensitivity, 43
Lighting, 53–56, 53*f*
Line of sight (LOS), 130, 144, 146–147, 172
LMR600, 156–160, 156*f*, 157*f*, 158*f*, 159*f*, 160*f*
 cable stripper, 158
 termination of, 159–160
Local area connection, 67, 68*f*
Local area network (LAN), 89, 96, 97–98, 283
Local recording, 50
Logarithms, 133
Loitering detection, 205
Long-range cameras, 48
LOS. *See* Line of sight
Low-light situations, 42, 44, 53
LPR. *See* License plate recognition
Lumens, 53
Luminous intensity, 53
Lux factor, 53, 54

M

M. *See* Make
MAC. *See* Media access control
Macroblocks, rasterordered, 35, 36
Main distribution facility (MDF), 5–6,
 183–185, 187, 209, 222
Make (M), 286
Manual focus, 45
Master book record (MBR), 249
MBR. *See* Master book record
MDF. *See* Main distribution facility
MDU. *See* Multi-dwelling units
Media access control (MAC), 92, 239
 address, 66–67, 92, 96–97, 128, 149
 address management, 126
 finding address of, 92, 92*f*
 sublayer, 126
Memory, 236–237
Mesh Networking, 145–147, 147*f*
 PtMP, 146–147
 PtP, 146
 seaf-healing of, 147, 147*f*
Mesh radio, 151–154
Mesh radio, test, PtP, 188–189
Metadata, 201, 203
Metcalfe, Robert M, 89
Microsoft Office, 192
Microsoft Project, 270
Microsoft Windows, 29, 67–68, 215, 237–238
 administrative tools, 238*b*
 chkdsk, 241
 device manager, 244, 244*f*
 NOS, 223
 peer-to-peer network for, 94, 94*f*
 PING tool of, 78–79, 104–105
 remote desktop, 211, 215, 215*f*, 241
 security update, 79
 troubleshooting hard drive in, 231*f*
 update, 121–122
Microwave RF band, 125–126, 126*f*, 130
Mid-infrared (mid-IR), 55
mid-IR. *See* Mid-infrared
Milliwatts (mW), 133
MIMO. *See* Multiple-input multiple-output
MJPEG. *See* Motion JPEG

Mobile devices, 33, 216, 216*f*, 292
Modular I/O units, 288
Monitoring, 4, 6, 7
  of potential target, 7
  queue length, 206
  speed, 207
  traffic flow, 206, 207
  VMS, 203
Monitors. *See also* Video displays
  CRT v. LCD, 21
  dual, 226, 227*f*
  flickering of, 23
  HDTV, 28*f*
  interlaced lines of CRT, 23
  LCD, 21
  multiple, 226, 227–228, 227*f*
  resolution of, 25, 26–27, 27*t*
  software video players v., 26*b*
Motherboard, 225, 234
  CPU v., 225, 226
  model number of, 225
  SATA ports on, 234*f*
Motion detection, 201, 206, 292
Motion JPEG (MJPEG), 32–33, 36–37
Motion Picture Expert Group (MPEG), 33–36
  compression, workings of, 34–36
  encoding path, 35*f*
  layers of, 35
  MPEG-1, 33, 114
  MPEG-2, 33, 114
  MPEG-3, 33
  MPEG-4, 33, 36–37, 114
  standards table, 34*t*
MPEG. *See* Motion Picture Expert Group
Multicast, 112–113, 112*f*, 121
  Class D IP address of, 112
  DVS, 112–113
  group, 114
  network traffic v., 112–113
  noise, 114
  packets, 113–115, 113*f*
Multi-dwelling units (MDU), 102
Multipath, 125, 143
Multiple element dipole antennas, 137.
    See also Dipole antennas

Multiple-input multiple-output (MIMO),
    125–126, 144
  802.11n, 131
  antennas, 144
  benefits of, 144
  diagram, 145*f*
  OFDM, 143
Multiplexing technology, 4, 144
mW. *See* Milliwatts

N
National Electrical Manufacturers Association
    (NEMA), 286–287
National Television System(s) Committee
    (NTSC), 9, 19, 22
Navy Pier, 13–15, 13*f*, 14*f*, 44, 197–199, 198*f*,
    205, 227*f*
NC. *See* Normally-closed
Near-infrared (near-IR), 55
near-IR. *See* Near-infrared
NEMA. *See* National Electrical Manufacturers
    Association
NETSTAT tool, 117
Netstumbler, 148
Network(s), 89
  accessibility, 239–241
  ad hoc self-contained, 127, 127*f*, 128
  adapter, 94
  anycasting on, 112
  bottlenecks, 107
  cabling for, 98–102
  delivery methods, digital video, 109–113
  design of, 93*t*, 95, 118
  diagrams, 261
  digital video, 108–109
  ethernet, 90, 93
  fiber optic, 100–101
  first electronic, 90*t*
  identifier, 104, 106
  infrastructure site survey, 183–187, 183*f*
  interface card, 92
  isolation of DVS, 185*b*
  latency, 162
  Layer 2, 118
  layer protocol, 103

location of, 274
multicasting on, 112–113, 112*f*
nodes of, 93*t*, 94
number, 103
peer-to-peer, 94–95, 94*f*
power of, 89, 89*b*
protocol of, 91
security, 118–119
sharing, 95*f*
small/medium company, 96*f*
TCP/IP, 90, 103
telephone prices v., 90*t*
token ring, 94, 95
traffic v. multicasting, 112–113
unicasting on, 109–111
unicasting v. bandwidth on, 111
VLAN, 107
wireless mesh, 145–147
Network management system (NMS),
      152–153, 167
   configuration of, 153*f*
Network operating system (NOS), 199–200,
      223, 237–239
Network traffic, 79
Networked video recorder (NVR), 195–196
Night vision, 48
NMS. *See* Network management system
NO. *See* Normally-open
No video, 78, 80*f*, 83*f*, 87–88, 116, 118, 162–163
Non-line-of-sight (nLOS), 144
Normally-closed (NC), 285–286, 287
Normally-open (NO), 285, 287
NOS. *See* Network operating system
NTSC. *See* National Television System(s)
      Committee
NVR. *See* Networked video recorder

O
Object classification, 208
Occupancy detection, 204–205
OCR. *See* Optical character recognition
Octets, 104, 104*f*, 106
OFDM. *See* Orthological frequency division
      multiplexing
Ohm meter, 305*b*

Omni-directional dipole antennas, 131–132
1 U components, 221–222
Open system Interconnection Reference
      model (OSI), 91, 91*f*, 96–97
Operating systems (OS), 90, 121, 195, 237–238
Operation Virtual Shield (OVS), 178
Optical character recognition (OCR), 173
Optics, 39, 40, 41, 50
Organizationally Unique Identifier (OUI), 113*f*
Orthological frequency division multiplexing
      (OFDM), 141–143, 142*t*
OS. *See* Operating systems
OSI. *See* Open system Interconnection
      Reference model
OVS. *See* Operation Virtual Shield

P
Packets
   broadcast, 113–115
   filtering, 240
   multicast, 113–115, 113*f*
   unicast, 113
PAL. *See* Phase alternating line
Pan-tilt-zoom camera (PTZ), 4, 9, 44,
      70–71, 162
   autofocus on, 45
   autotracking, 203, 204
   communication of, 62*b*, 81–84
   compatibility of, 81
   configuring communication for, 61
   control converters, 60–61
   controls sent through TCP, 108
   to digital encoder, connecting, 81
   event responses with, 289
   field of view, 183
   firmware, 71
   fixed camera v., 58–59
   flexibility of, 58*f*
   fps of, 49
   IP cameras, 50
   onscreen, 199, 200*f*
   protocols and communications of, 59–62
   troubleshooting, 71, 81–84, 83*f*
Parabolic dish antennas, 138
Passive systems, 1–2

Passwords, 68, 148
PATHPING tool, 119
PCB board, 88
PCI slot, 228–230, 229*f*
PCM. *See* Pulse code wave modulation
PCU. *See* Primary control unit
Peer-to-peer networks, 94–95, 94*f*
People counting, 205
Perimeter protection, 205
Personal identification number (PIN), 298
P-frames. *See* Forward predicted frames
Phase alternating line (PAL), 9, 19, 22
Photography, 40, 41–42, 53–56
Physical layer (PHY), 141
PIN. *See* Personal identification number
PING tool, 78–79, 104–105, 117, 119–121
Pixel(s), 21, 26–30, 27*t*
    analog camera/digital video encoder, 37
    analog television, 25
    bandwidth v., 31
    bits per, 26*t*
    color depth and, 25–26
    HDTV, 25
    of human recognition, 174, 175*f*, 176*f*, 176*t*
    LPR recommended, 176*t*
    mega, 27–31, 44–45, 48–49
    resolution v. facial, 176*t*
    upsampling, 26–27
Pixelation, 162
Plain old telephone service (POTS), 5–6, 10
Planning, 252–256, 267–268
    assumptions in, 254
    CPM of, 261–262, 262*f*, 262*t*
    design of, 254–256
    discovery step of, 253–254
    network diagrams of, 261
    process flow, 269*f*
    process improvement, 264, 264*f*
    project life cycle, 268
    project manager and, 256–267
    project stakeholders and, 268
    quality management, 263, 263*t*
    risk management, 264–267, 265*f*
    shipments and delivery, 256, 259*f*
    sponsors and, 268

    tasks of, 261–262, 256
    triple constraint of, 260–261, 260*f*
    WBS, 252–253, 256
PLC. *See* Programmable logic controllers
PMI. *See* Project Management Institute
PMO. *See* Project Management Office
PMP. *See* Project management professionals
PoE. *See* Power over Ethernet
Point-to-multipoint (PtMP), 146–147
Point-to-point (PtP), 146, 188–189
Polarization, 136
Poles, 12
Poor video, troubleshooting, 84–87, 86*f*,
        164, 165*f*
Port numbers, 109, 110*t*
POTS. *See* Plain old telephone service
Power, 124. *See also* Uninterrupted power
        supply
    camera, 51–52, 177–179
    data v., 178*b*
    failure of DC, 287
    gain and loss, 133–134
    grounding, 185–187
    law, 89
    output, 135*t*
    requirements, 51–52
    supply, 232
    surge, 79
    uninterrupted, 178
    VMS, 220
Power lines, MDU, 102
Power over Ethernet (PoE), 178, 297
Presurvey exploration, 187
Primary control unit (PCU), 4, 9
Primary lighting, 53
Print Screen, 29
Printing, 25, 27–31
Privacy laws, 7
Process improvement, 264
Processing power, 197*b*
Programmable logic controllers (PLC),
        286–287
Progressive scanning, 23–25, 24*f*
Project, 272–277. *See also* Planning
    agreement, 251–252

chain of command, 270*f*
change control process of, 271
closing process of, 277–280
execution process, 269–271
implementation, 251, 252*f*
initiating, 251–252
lifecycle, 268
manager, 256–267, 257*t*
payment for, 267–268
plan, 267–268
planning, 252–253
sponsors, 268
stakeholders, 268
status meetings and status reports, 270–271
tracking, 269–270
triple constraint of, 260–261, 260*f*
Project documentation deliverables, 272–277, 273*t*
as-built documents, 272–275
drawings, schematics, charts, and graphs, 275–276
status reports, 272
tracking, 276–277
version status and numbering of, 276
Project Management Institute (PMI), 180, 251–280
Project Management Office (PMO), 252*b*
Project management professionals (PMP), 251, 255*f*
PtMP. *See* Point-to-multipoint
PtP. *See* Point-to-point
PTZ. *See* Pan-tilt-zoom camera
Pulse code wave modulation (PCM), 20–21
Putty, 59
Pythagorean Theorem, 174, 174*f*

Q
QA. *See* Quality assurance
QC. *See* Quality control
QCIF. *See* Quarter Common Intermediate Format
QPSK. *See* Quadrature phase shift keying
Quadrature phase shift keying (QPSK), 143
Quality assurance (QA), 263, 263*t*
Quality control (QC), 263, 263*t*

Quantization, 35
Quarter Common Intermediate Format (QCIF), 31
Queue length monitoring, 206
QuickTime, 31–32

R
Rack mount equipment, 185, 221, 221*f*, 222
RAD. *See* Rapid application development
Radiation, 126*f*
Radiation patterns, 135, 135*f*, 136, 137, 137*f*, 138*f*, 139*f*
Radio(s), 100, 134–135, 134*f*
configuration of mesh, 151–154
ESSID settings for, 154
mesh, 146, 147, 163
mobile, 163–164
test, PtP mesh, 188–189
testing of wireless, 279
Radio frequency (RF), 3–4, 19, 71, 123, 124–126
AC of, 124
amplifier, 124–125
bandwidth sharing of, 123
barriers, 129, 129*t*
carrier frequencies of, 100–101
digital v., 9
distortion of, 126
FCC regulation of, 20–21, 125, 126*b*, 130, 132
fiber optics v., 100–101
gain and loss, 133–134
interference, 129–130
ISM, 140
measurements of, 125, 125*b*
microwave band of, 125–126
multipath, 125
non-licensed, 150
rain v., 125
reflection, 125
spectrum, 123, 187, 188, 188*f*
terrestrial broadcasting, 143
wattage of unlicensed spectrum of, 133
wireless communications and, 123
WLAN application, 125–126, 126*f*

RAID. *See* Redundant array of independent disks
RAM. *See* Random access memory
Random access memory (RAM), 236–237
Rapid application development (RAD), 203
Rasterized images, 275, 275*f*
RCA composite connector, 71
Real-time data retrieval, 204
Real-time streaming protocol (RTSP), 114
Real-time transport protocol (RTP), 114
   communication between host computer and IP camera, 115*f*
   workings of, 114*f*
Reboot, 164*b*, 167
Received signal strength indication (RSSI), 189
Red, green, and blue (RBG), 21, 29
Redundant array of independent disks (RAID), 235–236, 236*f*, 236*t*, 245, 274
Reflection, RF, 125
Refraction, 39–40
Relays. *See* Electronic relay connections
Remote access, 115, 211–215
Remote desktop, 211, 215, 215*f*, 241
Remote reboot, 52
Remote viewing, DVS, 240–241
Request-to-send/Clear-to-send (RTS/CTS), 128–129
Resolution, 25, 28*f*
   2CIF, 31
   4CIF, 29*f*, 31
   16CIF, 31
   CIF, 31, 34
   digital video surveillance, 26–27, 27*t*, 28*f*, 31
   facial pixels v., 177*t*
   of human recognition, 174
   maximizing display, 29
   megapixel, 59
   of monitors, 25, 26–27, 27*t*
   progressive scanning and, 23
   QCIF, 31
   storage v., 34
   of video games, 71–72
   worldwide standards of, 22, 22*t*
RF. *See* Radio frequency
RGB. *See* Red, green, and blue

Risk, 264–267, 265*f*
RJ45 boots, 75, 75*f*
RJ45 connectors, 75, 98*f*
RJ45 port, 45, 63
Router, 96, 107
RSSI. *See* Received signal strength indication
RTP. *See* Real-time transport protocol
RTS/CTS. *See* Request-to-send/Clear-to-send
RTSP. *See* Real-time streaming protocol

S
S2 Netbox Solution, 298, 299*f*
S3. *See* Smart Surveillance Solution
Salient motion detection, 207–208
SAN. *See* Storage area networks
SATA. *See* Serial AT attachment
Scalability, 239
Scalable link interface (SLI), 226–227
Screen capture, 29–30
SDK. *See* Software development kit
SECAM. *See* Soviet Union Sequential Couleur avec Memoire
Secondary lighting, 53
Sector antennas, 138, 139*f*
Security, 11–12
   false sense of, 177–178
   integration, 281–294
   management and monitoring, centralized, 283
   rules for digital, 148
   rules for wireless, 148–150
   settings, 154, 154*f*
   VMS, 222
   wireless, 148–149
Security camera systems, 11*f*
   4CIF high-resolution analog video, 29*f*
   active, 1–2
   analog v. digital v. megapixel, 48–50
   cost of, 1
   costs of, 45, 46*t*
   DHCP v., 66–67
   disabling, 11
   economical, 46
   fake, 3*b*
   formats of, 33

high-resolution, 26–27
IR, 55–56
low-resolution, 26–27
passive, 1–2
power requirements of, 51–52
protection of, 11, 12
remote reboot of, 52
sensors of, 43
wireless everything, 149*b*
Selenium photocell, 40
Self-healing, 147, 147*f*
Serial AT attachment (SATA), 228, 232,
 233*f*, 234*f*
Serial communication protocols, 81–84
Serial port, RS232, 63, 66*f*
Serial web interface, 61–62, 61*f*
Server
 scalability of, 239
 system requirements, 224*t*
 workstation v., 222–223
Service set identifier (SSID), 127–128, 149
Sharing, network, 95*f*
Sharpen, 30
Shipping and receiving, 185*b*
Shortwave frequency spectrum, 101–102
Shutter speed, 40, 41–42, 42*f*, 50–51
Signal loss, 77, 77*f*, 156
Simple mail transport protocol (SMTP),
 97–98
Single name identifier (SSID), 146
Single Pole Double Throw (SPDT), 286
Single Pole Single Throw (SPST), 286
Single-lens reflex (SLR), 39
Site surveys, DVS, 171
 camera location of, 181–183
 camera/video, 180–183
 final, 180
 form, 181
 formal agreement of, 180
 IDF, 184–185
 network infrastructure, 183–187, 183*f*
 power requirement of, 177–179
 process of, 191*f*
 teams, 181
 wireless, 187–193

16CIF. *See* Sixteen times Common
 Intermediate Format
Sixteen times Common Intermediate Format
 (16CIF), 31
Smart Surveillance Solution (S3), 203, 207
Smart Vault, 196, 303*f*
Smith Museum of Stained Glass, 197–199, 289,
 290, 291, 292*f*
Software, 39
 applications, devices v., 202*t*
 database, 201
 digital encoder, 63
 drivers, 200
 hardware v., 26*b*, 39
 ipConfigure, 196
 malicious, 240
 network management, 152–153
 plug-in, 115
 port numbers, 109
 remote reboot v., 52
 tools, 190
 troubleshooting, 79, 117
 Verint SConfigurator, 65–66, 65*f*,
  66*f*, 121
 video management, 195–218
 video surveillance, 26–27
 VM, 116
 VMS, 9, 10, 29, 60, 121
 wireless propagation, 192–193, 192*f*
Software development kit (SDK), 281
Soviet Union Sequential Couleur avec
 Memoire (SECAM), 22
SPDT. *See* Single Pole Double Throw
Spectrum analysis, RF, 188
Spectrum analyzer, 187, 188*f*
Speed monitoring, 207
SPI. *See* Stateful packet inspection
Sponsors, 268
SPST. *See* Single Pole Single Throw
Spyware, 6–7
SSID. *See* Service set identifier
Stateful packet inspection (SPI), 97
Storage, 34, 50, 211, 220–222, 289
Storage area networks (SAN), 211, 219,
 281–282

Streaming video, 188. *See also* Real-time
    streaming protocol
    bandwidth v., 107
    CSMA/CA v., 140
    multiple streams of, 34
    TCP and, 108
    UDP and, 108
    unicast, 109–111
Subnet address, 104
Subnet masks, 67–68, 104, 106
Subsampling, 35
Surge protectors and suppressers,
    178–179
Surveillance, 171. *See also* Digital video
    security
    analog television and, 4
    audio, 63
    capable guardian in, 2–3
    crime v., 1–2
    detection in, 3
    deterrence in, 1–2, 3
    efficiency in, 2
    fake video cameras in, 3*b*
    formats, video, 31–33, 31*t*
    hardware, evolution of video, 39
    history of visual, 1
    programming of, 2–3
    software, 26–27
    synchronous, 6
    theory of visual, 1
Switches, 96–97, 107, 118
    AP, 126–131
    dip, 81, 84*f*
    of star network, 103
    stupid, 118
    testing of, 278–279
    VLAN, 107
Synchronization, 6, 114

T

TAC. *See* Technical Advisory Committee
TCI/IP. *See* Transport control protocol/
    Internet Protocol
TCP. *See* Transmission control protocol
Technical Advisory Committee (TAC), 140

Technology, evolution of, 6, 56
Telecommunication technologies, 7
Television. *See also* Closed circuit television
    (CCTV)
    analog, 4, 71
    color, 22
    color depth of analog, 21
    high definition, 8
    resolution of, 25
Television Bureau of Advertising, 19–20
Temperature, 57*b*
Terminal services, 241
Thermal cameras, 48
Thermal technology, 55–56
Thermal-IR, 55
    IR illumination v., 55
Token ring topology, 94, 95
Tools. *See also* Laptop
    communication, 193
    crimping, 157*f*, 159
    digital camera as, 193, 193*b*
    disk management, 247–248, 248*f*
    hardware, 189–193
    lens calculator, 174, 175*f*
    troubleshooting, 163*b*
    wireless site survey, 187, 189–193
Topology, 95, 96*f*
    bus, 95
    CCTV to DVS conversion, 10*f*, 184*f*
    EAC, 296*f*, 300, 301*f*
    ethernet, 93, 93*t*
    IP-based, 297, 298*f*
    mesh networking, 147*f*
    star and tree, 94, 95, 103–106
    token ring, 94, 95
    wireless AP, 127*f*
TRACERT tool, 117
Traffic flow monitoring, 206, 207
Transfer control protocol (TCP), 91
    port numbers, 110*t*
    PTZ controls sent through, 108
    reliability of, 108
    synchronicity of, 108
Transistor-to-transistor logic (TTL), 287
Transmission control protocol (TCP), 108

network security and, 118–119
streaming video and, 108
transmission process of, 108
Transport control protocol/Internet
        Protocol (TCI/IP), 69f, 90, 91,
        103, 208b
    application layer of, 91
    foundation of, 108
    Internet layer of, 91
    IP address assignment for, 103
    link layer of, 108
    network access layer of, 92, 108
    transport layer of, 91
Troubleshooting
    analog cameras, 78–80, 80f, 82f
    camera power, 87f
    digital video cables, 79
    digital video connectors, 79
    DVR card, 242–249, 243f
    DVS, 77–88, 80f, 161, 162
    equipment, 163b
    ethernet networks, 119
    hard drive, 231f, 245–249, 246f
    incompatibilities, 119
    I/O connections, 302–305, 306f
    IP camera, 81, 82f, 83f
    IP networking, 119–121, 120f
    laptop, 116
    multidimensional, 116f
    NETSTAT tool for, 117
    no video, 83f, 118
    PATHPING tool for, 119
    physical wire connections, 304
    PING tool for, 117, 119–121
    poor video, 84–87, 86f, 164, 165f
    PTZ, 81–84, 83f
    software, 79, 117
    tools, 163b
    TRACERT tool for, 117
    VMS, 79, 116, 216–218, 217f
    what usually goes wrong, 117–118
    with wireless mobile unit, 163
    WLAN, 160–167
TTL. See Transistor-to-transistor logic
Twisted-pair copper cable, 98t

U
UDP. See User datagram protocol
Unicast, 109–111, 111f, 114
    bandwidth v., 111
    packets, 113
Uniform resource locator (URL),
        104–105
UNII. See Unlicensed National Information
        Infrastructure (UNII)
Uninterrupted power supply (UPS), 179
    backup batteries of, 179
    Standby, 179
    VA rating of, 179
Unit discovery, 63
Unlicensed National Information
        Infrastructure (UNII), 140, 141
Unshielded twisted pairs (UTP), 93, 98,
        98f, 98t
Upconverting, 27–31
Upsampling, 26–27, 29b, 30f, 59
    convert for print in, 30
    full screen display in, 29
    low-light, 53
    maximize display resolution in, 29
    save image in, 30
    screen capture in, 29–30
Urban Eye, 1–2
URL. See Uniform resource locator
User access, 210–211
User datagram protocol (UDP), 91,
        108–109
    asynchronicity of, 108–109
    multicasting through, 112–113
    network security and, 118–119
    port numbers, 110t
    RTP and, 114
    streaming video and, 108–109
    unreliability of, 108–109
User groups, 210–211
    administrators, 210
    archive player users, 211
    live view users, 211
    power users, 211
User management, 208–211
UTP. See Unshielded twisted pairs

V

VA. *See* Volt-ampere

Vector graphics, 275–276, 275*f*

Verint SConfigurator software, 65–66, 65*f*, 66*f*, 121

VGA. *See* Video graphics array

Video analytics, 2–3, 202–207, 282–283, 289

  abandoned object detection, 207

  area obstruction detection, 205

  camera alignment detection, 206

  camera tampering alarm, 206, 293–294

  facial detection, 206

  fall detection, 206

  intelligent screen verification, 205

  loitering detection, 205

  object classification, 208

  occupancy detection, 204–205

  panic alarm, 207

  people counting, 205

  perimeter protection, 205

  PTZ autotracking, 203, 204

  queue length monitoring, 206

  real-time data retrieval, 204

  salient motion detection, 207–208

  software, 203

  speed monitoring, 207

  traffic flow monitoring, 206

  VMS integration, 204

Video Balun transceivers, 5–6

Video cameras

  costs of, 46*t*

  economical, 46

  evolution of, 39

  fake, 3*b*

  fixed digital IP, 36

  IP, 45

  low-light environments and, 42

  PTZ, 4, 36, 44

  still camera v., 42

  workings of, 39–45

Video compression formats, 7, 31–32

Video conferencing, 31, 114

Video displays, 71

Video footage, monitoring of live and archived, 2

Video graphics array (VGA), 71–72, 74

Video graphics card, 226–227, 227*b*

Video management system (VMS), 9, 195–218, 202*f*, 237–238, 281, 289. *See also* Video analytics

  adding cameras to, 208

  alarm and event management on, 288

  application, 195

  applications v. devices, 202*t*

  archiving and storage of, 211

  audio compatibility with, 63

  blinking of, 162

  compatibility of, 60, 81, 199–201

  digital encoder v., 78

  drivers of, 60

  DVR card compatibility with, 232

  DVS requirements for, 219–222

  ethernet network of, 9

  event responses of, 289–292

  hardware system requirements, 224–225, 224*t*

  high-end, 283

  implementation of, 197, 199

  integration, 204

  interface with map, 200*f*

  IP cameras v., 50

  log files of, 210

  market share of, 201

  metadata and database, 201

  mobile, 216

  monitoring with, 203

  power of, 220

  providers, 196

  PTZ communication with, 59–60

  rack mount, 221, 221*f*, 222

  remote access, 211–215

  remote access of, 115

  S3, 203

  security, 222

  software, 9, 10, 29, 121

  troubleshooting, 77, 216–218, 217*f*

  user access, 210–211

  user management, 208–211

Video motion detection (VMD), 292

Video players, monitors v. software, 26*b*

Video processing, 197*b*

Video standards, worldwide, 22, 22*t*

Video surveillance appliance (VSA), 196
  ipConfigure, 196

Video surveillance hardware
  economical, 46–47
  evolution of, 39
  malfunction of, 52–53
  night vision, 48
  professional, 47, 58, 60–61
  specialty, 47–48
  temperature and climate control of, 57*b*

Virtual local area network (VLAN), 9, 96–97,
    107–108

Virtual machine (VM), 116

Virtual PC, 116

Virtual private network (VPN), 215, 216, 283

Virus, 235, 240

VLAN. *See* Virtual local area network

VM. *See* Virtual machine

VMD. *See* Video motion detection

VMS. *See* Video management system

Voiceover IP (VoIP), 8, 10

VoIP. *See* Voiceover IP

Voltage, 178–179, 186–187

Voltage standing wave ration (VSWR), 126

Volt-ampere (VA), 179

VPN. *See* Virtual private network

VSA. *See* Video surveillance appliance

VSIP port, 121

VSWR. *See* Voltage standing wave ration

W

W. *See* Watts

Waiting for signal, 162

WAN. *See* Wide-area network

Water, 57

Watts (W), 133

Wavelengths, 55, 125*b*

WBS. *See* Work breakdown structure

Weather, 56–58

Wide-area network (WAN), 96, 97–98

Windows Media Audio, 31–32

Windows Media Player, 31–32

Windows Media Video, 31–32

Wired, 91–92

Wireless antenna, 155–160, 155*f*

Wireless communications, 123, 124

Wireless everything cameras, 149*b*

Wireless local area network (WLAN), 123–124
  channel planning for, 150
  configuring AP radios for, 150–151
  connectivity of, 164, 166*f*
  802.11, 138–140
  802.11a, 141
  802.11b, 140–141
  802.11g, 143
  802.11n, 143–144
  802.11s-mesh, 145–146
  power calculations of, 132
  RF spectrum used by, 125–126, 126*f*
  SSID of, 127–128
  standards, 138–147, 142*t*
  troubleshooting, 160–167

Wireless mesh networking, 145–147

Wireless mobile unit, 163, 188, 189*f*

Wireless security, 148–150

Wireless site survey, 187–193
  flowchart, 191*f*
  presurvey exploration, 187
  PtP mesh radio test of, 188–189
  results of, 189
  RF spectrum analysis of, 188
  tools for, 187, 189–193

WLAN. *See* Wireless local area network

Work breakdown structure (WBS), 252–253,
    256, 257*t*

Workstation
  SATA ports of, 233, 233*f*, 234*f*
  server v, 222–223
  system requirements, 224*t*

Y

Yagi antennas, 137, 137*f*